T0340377

Ethics, CSR and Sustainability (ECSRS) Education in the Middle East and North Africa (MENA) Region

The Middle East and North Africa (MENA) region is undergoing significant socio-political and developmental transition. Although interest in corporate social responsibility (CSR) in the region is growing, little research has addressed corporate social responsibility education and its potential impact. CSR has an important role to play in the socio-economic development of the Middle East and North Africa due to the volatility and developmental needs of the region. Recent research has highlighted that the vitality of the institutional environment and the needs of multiple stakeholders in CSR are not necessarily consistent with the notion of CSR in the West.

This book compiles conceptual, contextual, and empirical research that addresses the concepts of CSR, ethics, and sustainability education in the MENA region, with a special emphasis on how educators can bridge to the Giving Voice to Values approach.

This book presents a much-needed portfolio of articles from authors based in Egypt, Morocco, the Sultanate of Oman, Qatar, and the United Arab Emirates (UAE), highlighting first an overview of the topic and its corresponding publications in the MENA region, then presenting several exemplary cases related to ECSRS application in various countries.

Noha El-Bassiouny is Vice Dean for Academic Affairs, Professor and Head of Marketing at the Faculty of Management Technology, the German University in Cairo (GUC), Egypt. She also acts as the academic coordinator of the Business and Society research group, which aims at bridging the interface between business and society in the modern world in terms of research, teaching, and community outreach. Her research interests lie in the domains of consumer psychology, Islamic marketing, ethical marketing, corporate social responsibility, and sustainability. She has wide international exposure and has published her works in reputable journals including the *Journal of Business Research*, the *International Journal of Consumer Studies*, the *Journal of Consumer Marketing*, the *Social Responsibility Journal*, the *Journal of Islamic Marketing*, the *International Journal of Bank Marketing*, the *International Journal of Pharmaceutical and Healthcare Marketing*, the *Journal of Cleaner Production* as well as *Young Consumers*. She is currently the Associate Editor of the *Journal of*

Islamic Marketing. She has also received many international awards including the prestigious Abdul Hameed Shoman Arab Researchers Award (2019) as well as several Emerald Outstanding Reviewer Awards and Highly Commended Paper awards.

Dina El-Bassiouny is an assistant professor in the Accounting Department at Zayed University, Abu Dhabi, UAE. Previously, she has taught undergraduate courses in accounting at a number of institutions, including the American University in Cairo, Misr International University and Future University, Egypt. She holds a PhD in business and economics, concentration in sustainability and corporate social responsibility, from RWTH Aachen University, Germany, and an MSc in accounting from the German University in Cairo. Her research interests include business ethics, corporate social responsibility, sustainability reporting, and corporate governance. She has published her works in reputable outlets including *Social Responsibility Journal (SRJ)*, *Sustainability Accounting, Management and Policy (SAMP) Journal*, *Management of Environmental Quality (MEQ)*, and the *American Accounting Association (AAA)* Conference.

Ehab K. A. Mohamed is the Vice President for Students Affairs at the German University in Cairo. He is also the Dean and Professor of Accounting at the Faculty of Management Technology. Prior to joining GUC he worked for 10 years at Sultan Qaboos University, Oman. He graduated from Cairo University and received his MSc and PhD from Cass Business School, London. He is a Fellow of the Chartered Institute of Internal Auditors, UK. His areas of research are in auditing, fraud, performance measurement, business education, financial reporting, and corporate governance. He has published a number of papers in international refereed journals and presented papers at numerous international conferences. He published two books on financial accounting and auditing.

Mohamed A. K. Basuony is Associate Professor of Accounting at the School of Business at The American University in Cairo. Prior to that, he worked at the German University in Cairo and Faculty of Commerce, Ain Shams University. He received both his bachelor with honors and master's degrees from the same university. He obtained his PhD degree from Brunel University in the UK. His work experience includes 20 years teaching undergraduates as well as postgraduates in Egypt and the UK. His areas of research are in performance management, balanced scorecard, management and strategic control, corporate governance, and corporate social responsibility.

Giving Voice to Values
Series Editor: Mary C. Gentile

The *Giving Voice to Values* series is a collection of books on Business Ethics and Corporate Social Responsibility that brings a practical, solutions-oriented, skill-building approach to the salient questions of values-driven leadership.

Giving Voice to Values (GVV: www.GivingVoiceToValues.org) – the curriculum, the pedagogy and the research upon which it is based – was designed to transform the foundational assumptions upon which the teaching of business ethics is based, and importantly, to equip future business leaders to know not only what is right – but how to make it happen.

Giving Voice to Values in the Legal Profession
Carolyn Plump

Giving Voice to Values in Accounting
Tara J. Shawver and William F. Miller

Giving Voice to Values as a Professional Physician
Ira Bedzow

Authentic Excellence
R. Kelly Crace and Robert L. Crace

Ethics, CSR and Sustainability (ECSRS) Education in the Middle East and North Africa (MENA) Region
Edited by Noha El-Bassiouny, Dina El-Bassiouny, Ehab K. A. Mohamed, and Mohamed A. K. Basuony

Ethics, CSR and Sustainability (ECSRS) Education in the Middle East and North Africa (MENA) Region

Conceptualization, Contextualization, and Empirical Evidence

Edited by Noha El-Bassiouny, Dina El-Bassiouny, Ehab K. A. Mohamed, and Mohamed A. K. Basuony

LONDON AND NEW YORK

First published 2020
by Routledge
2 Park Square, Milton Park, Abingdon, Oxon OX14 4RN

and by Routledge
52 Vanderbilt Avenue, New York, NY 10017

Routledge is an imprint of the Taylor & Francis Group, an informa business

British Library Cataloguing-in-Publication Data
A catalogue record for this book is available from the British Library

Library of Congress Cataloging-in-Publication Data
A catalog record has been requested for this book

ISBN: 978-0-367-90125-7 (hbk)
ISBN: 978-0-367-51511-9 (pbk)
ISBN: 978-1-003-02276-3 (ebk)

Typeset in Bembo
by Wearset Ltd, Boldon, Tyne and Wear

Contents

Notes on contributors ix
Foreword xiii
Editorial Review Board xiv

Introduction 1

1 **A citation-based systematic literature review on ECSRS education in MENA region countries** 8
NISAR AHMAD, HELENA H. KNIGHT AND IRFAN BUTT

2 **Corporate social responsibility (CSR) and the developing world: highlights on the Egyptian case with implications for CSR education** 34
DINA EL-BASSIOUNY

3 **Approaching the Giving Voice to Values (GVV) pedagogy in business ethics education: the case of the business ethics course at the German University in Cairo (GUC), Egypt** 51
NOHA EL-BASSIOUNY, SARA HAMED, NESMA AMMAR, HADEER HAMMAD AND HAGAR ADIB

4 **Education for sustainable development: a means for infusing social responsibility in higher education in Egypt** 69
HEBA EL-DEGHAIDY

5 **Business ethics in the business schools in Morocco** 92
WAFA EL-GARAH AND ASMAE EL MAHDI

6 **Integrating sustainability and CSR concepts in the College of Business and Economics (CBE) curriculum: an experiential learning approach** 127
DALIA ABDELRAHMAN FARRAG AND SHATHA OBEIDAT

Conclusion 145

Index 148

Contributors

Hagar Adib has a PhD in marketing from the German University in Cairo. Hagar has developed a passion for interdisciplinary approaches to overcome unsustainable consumption patterns. She has several research interests including business ethics, sustainability, pro-environmental behavior, green innovations, and materialism.

Nisar Ahmad received his master's and PhD from Aarhus University Denmark and worked in the University of Southern Denmark as Assistant and Associate Professor. Currently, he is working as Assistant Professor at the Department of Economics and Finance, Sultan Qaboos University. Recently, he has published in the *Journal of Macroeconomics*, *International Migration* and *Journal of Borderland Studies*.

Nesma Ammar has her PhD in marketing from the German University in Cairo (GUC), Egypt. Her research interests include consumer psychology, sustainable consumption, young consumers, consumerism, ethics, and education. She is interested in researching the new trend of social marketing and its role in enhancing consumers' well-being, the impact of materialism on consumer values, and bridging the gap between theory and practice. She has attended several international conferences and participated in different research projects. Her recent publications include Ammar, N., El-Bassiouny, N., & Hawash, R. (2016). Materialism and healthy food consumption: Can health education play a role? *Social Business*, 6(4), 377–401.

Irfan Butt received his PhD from Sprott School of Business, Carleton University, Canada, and MBA from the Thunderbird School of Global Management, Arizona State University, USA. He is currently based at Lakehead University, Thunder Bay, Canada. Prior to that, he has taught at the Sultan Qaboos University, Oman, and Lahore University of Management Sciences, Pakistan. His research has been published or accepted for publication in the *International Journal of Operations and Production Management*, *Journal of Travel and Tourism Marketing*, *Journal of Consumer Marketing*, *International Journal of Technology Management*, *Journal of Global Business Advancement*, and *Journal of Teaching in International Business*.

Dina El-Bassiouny is an Assistant Professor in the Accounting Department at Zayed University, Abu Dhabi, UAE. Previously, she has taught undergraduate courses in accounting at a number of institutions, including the American University in Cairo, Misr International University and Future University. She holds a PhD in business and economics, concentration in sustainability and corporate social responsibility, from RWTH Aachen University, Germany, and an MSc in accounting from the German University in Cairo. Her research interests include business ethics, corporate social responsibility, sustainability reporting and corporate governance. She has published her works in reputable outlets including *Social Responsibility Journal (SRJ)*, *Sustainability Accounting, Management and Policy (SAMP) Journal*, and the *American Accounting Association (AAA)* Conference.

Noha El-Bassiouny is Vice Dean for Academic Affairs and Professor and Head of Marketing at the Faculty of Management Technology, the German University in Cairo (GUC), Egypt. She also acts as the academic coordinator of the Business and Society research group, which aims at bridging the interface between business and society in the modern world in terms of research, teaching, and community outreach. Her research interests lie in the domains of consumer psychology, Islamic marketing, ethical marketing, corporate social responsibility, and sustainability. She has wide international exposure and has published her works in reputable journals including the *Journal of Business Research*, the *International Journal of Consumer Studies*, the *Journal of Consumer Marketing*, the *Social Responsibility Journal*, the *Journal of Islamic Marketing*, the *International Journal of Bank Marketing*, the *International Journal of Pharmaceutical and Healthcare Marketing*, *Journal of Cleaner Production* as well as *Young Consumers*. She is currently the Associate Editor of *the Journal of Islamic Marketing*. She has also received many international awards including Emerald Outstanding Reviewer Awards and several Highly Commended Paper awards.

Heba El-Deghaidy is currently chair of the Department of International and Comparative Education at the Graduate School of Education. El-Deghaidy is an Associate Professor of Science Education and leads the STEAM education initiative as an international approach to an interdisciplinary learning model. Her doctoral degree in science education comes from the University of Birmingham, UK. She was the principal investigator of the bilingual STEAM education project funded by British Gas. El-Deghaidy is the Co-PI for an Erasmus-funded project called School and University Partnership for Peer Communities of Learners (SUP4PCL). El-Deghaidy's research focuses on finding a means to provide for quality teaching and learning opportunities. Her research interests focus on teacher education in general and science teachers in particular. She has publications in the areas of STEM/STEAM education, education for sustainable development (ESD), using educational technology in teaching and learning, teaching portfolios, continuing professional development and other areas.

Wafa El-Garah is Associate Professor of Management Information Systems (MIS) and Project Director of the Center for Business Ethics at Al Akhawayn University in Ifrane (AUI). She served as Vice President for Academic Affairs (VPAA) at AUI, where she successfully led the university through the last stage of the NEASC accreditation process. She served as Dean of the School of Business Administration for five years. She led the school through a successful international EPAS reaccreditation process of the bachelor program. Dr. El Garah holds a PhD in MIS from the University of Central Florida, USA, and an MBA in marketing and management from the University of North Texas, USA. She serves on the board of the international accreditation agency EFMD. She is also a member of the EPAS Committee. Her research has appeared in outlets such as *Data Base for Advances in Information Systems, Communications of the AIS* as well as numerous international conference proceedings.

Asmae El Mahdi is a Program Manager at the School of Science and Engineering at Al Akhawayn University. She is also Project Manager at Al Akhawayn Center for Business Ethics. Asmae has extensive experience and an excellent track record in e-government projects. Her expertise is in the use of information and communication technologies for good governance. She holds a bachelor of arts degree in International Studies from Al Akhawayn University. Her research interests include: business ethics, corporate social responsibility (CSR), ethical practice and compliance standards in organizations, and localizing and mainstreaming Giving Voice to Values (GVV) in the curricula of corporates and business schools.

Dalia Abdelrahman Farrag is currently an Associate Professor of Marketing at the College of Business and Economics – Qatar University. She has been awarded the Fellow status from the Chartered Institute of Marketing (CIM) – UK, and obtained her PhD from Alexandria University with a joint supervision from the University of Utah, David Eccles School of Business, USA. Her teaching experience includes principles of marketing, marketing management, marketing research, consumer behavior, advertising management, marketing channels, international marketing and advanced marketing at both the undergraduate and postgraduate levels. Dr. Farrag's research interests include interpretive consumer research (ethnography), consumer behavior, Islamic marketing, political marketing, entrepreneurship, CSR, branding, and shopping behavior. Dr. Farrag has also received several international research awards like Emerald Most Outstanding Paper Award in 2015, amongst others.

Hadeer Hammad has a PhD in marketing from the German University in Cairo. Her research interests include responsible marketing, sustainable consumption, social responsibility, consumer psychology, luxury consumption, and materialism. She is interested in researching the moral aspects of the marketing field, the merge between business and societal goals, the impact of materialistic values on consumption habits and consumer well-being, and the interplay between consumer psychology and consumer behavior.

Sara Hamed is Assistant Professor in Marketing at the Faculty of Management Technology at the German University in Cairo (GUC), Egypt. She obtained her PhD from the GUC in 2016. Her research interests include the integration between the fields of marketing and design, sustainability and well-being concerns, as well as digital marketing. Dr. Hamed teaches a range of marketing and research courses and was involved in finance courses. In her 10 years of teaching, Dr. Hamed has focused on establishing an interactive teaching/education environment for her students. She has organized a cross-faculty workshop (between Marketing and Design) and was involved in a Microfinance research group at the university aimed at giving students hands-on experience from the field of practice and research. Dr. Hamed also supervised the Google Online Marketing Challenge from 2010 until its closure in 2017 and still focuses on teaching her students the latest in the field of digital marketing.

Helena Knight is an Assistant Professor in the Marketing Department at the College of Economics and Political Science. She received her PhD in Marketing and Strategy, Cardiff University, UK, 2015. She has worked with numerous high-profile business corporations in the UK, including FTSE100 companies and nonprofit organizations operating in the arts context. Helena has also served as a reviewer of submissions for the *Journal of Business Research* and co-organized numerous symposia, seminars and workshops as a founding member of a research group at Cardiff University.

Shatha Obeidat is an Assistant Professor of Management in the College of Business and Economics at Qatar University, Qatar. Dr. Obeidat received her PhD from the University of Newcastle, Australia. She has published in reputable international journals, including *Human Resource Management*, *Personnel Review*, and *Employee Relations*. Her research interests focus on strategic human resource management, e-HRM, green HRM, and corporate social responsibility.

Foreword

Sustainable management: the new business paradigm in rising economies

In rising economies there is increased awareness for sustainability and ethics as a prerequisite of business success. Multinational enterprises and local companies alike have an important role for the sustainable development of a society. Modern businesses consider social and environmental factors with the goal of positive impact on society while achieving business success. This responsibility is not a burden but an opportunity to align economic and social success.

As business schools, we need to educate the future leaders and prepare them for problem-solving in complex global and local economies. In this publication, the editors show interesting insights into the new understanding of management as well as business application in the MENA region. This is not only interesting for businesses in this area, but also a good source of innovation for other regions.

I would personally like to thank Noha El-Bassiouny and her team as well as all the authors of this publication for sharing their knowledge and latest achievements in the field of sustainable management and responsible leadership education. These examples not only show the way forward for the MENA region, but also help to develop a global mindset of sustainability. Together we make sustainability the new normal!

Prof. Dr. René Schmidpeter
General Secretary of the World Institute for Sustainability and Ethics (WISE)
Director of the Center for Advanced Sustainable Management (CASM)

Editorial Review Board

REVIEW BOARD TABLE

Introduction

Corporate social responsibility (CSR) has an important role to play in the socio-economic development of the MENA region given its high volatility and developmental needs (Al-Abdin et al., 2018). Recent literature has highlighted, however, that the vitality of the institutional environment and the needs of multiple stakeholders in the CSR ecosystem varies in the Middle East (Al-Abdin et al., 2018; Jamali & Neville, 2011), and that the Middle East has a specific understanding of CSR that is not necessarily consistent with the notion of CSR in the West (Jamali & Sidani, 2012).

Apart from this variety in CSR needs and in the interaction between business and societal stakeholders in the region, CSR remains a vital and emerging research domain in developing countries in general (Jamali & Karam, 2016), and in the Middle East region specifically (Al-Abdin et al., 2018). Despite the multiplicity of studies showing the potential impact of CSR and the variety of approaches utilized by MNCs and SMEs in the MENA region (for a review see Al-Abdin et al., 2018), studies on CSR education in the region remain almost non-existent.

Although studies have documented the potential impact of demographic factors, such as religiosity and education (Cheah et al., 2011; Jamali & Sidani, 2013), on CSR managerial attitudes, little research exists on the educational preparation that these potential managers get. Recently, the new and emerging concept and methodology of the Giving Voice to Values (GVV) approach has made global presence (see, e.g., Arce & Gentile, 2015). The GVV approach highlights that ethics education needs to be positive and starts on the premise that people in organizational settings know the right thing to do and want to act upon it, i.e., it is post-decision-making (Gentile, 2012). In addition to the fact that there is almost no literature on CSR education in the MENA region, bridging with the GVV approach is non-existent to date.

It is safe to say that technological advancements have, unfortunately, not been without cost. Globalization coupled with industrialization has been the cause of huge environmental damage. This is when "sustainable development," as a term, was coined. Generally, to reach such development, massive human contributions in all fields are required, including management and its body of scholarship.

Corporations have been increasingly catering to their social responsibilities and disclosing more complex social and environmental information. Responsively, research on corporate social and environmental disclosure has increased as well (Gray et al., 2001). Several theories were used to explain this interesting phenomenon including stakeholder and/or legitimacy theories (see, for example, Boesso & Kumar, 2007; Campbell, 2004; Clarkson et al., 2007; Magness, 2006; Liu & Anbumozhi, 2009), agency theory (Gray et al., 2001; Meek et al., 1995) and business ethics theory (see, for example, Cho et al., 2006).

Prior studies have attempted to understand, analyze, and interpret the determinants of such corporate social responsibility (CSR) activities and their manifestations in the form of disclosure. The effect of one or more different factors on the quantity and quality of environmental information disclosed by corporations has been tested. This includes regulations and standards (see, for example, Jorgensen and Soderstom, 2006), competitive strategy employed by the firm (see, for example, Ling, 2007), the strategic posture of the firm (see, for example, Van der Laan, 2009; Magness, 2006), company size and location (see, for example, Barbu et al., 2012), degree of corporate political activity (see, for example, Cho et al., 2006), multinationality (see, for example, Meek et al., 1995), the degree of media exposure (see, for example, Aerts & Cormier, 2009) and others. One of the important influencing factors is, in fact, managers' environmental knowledge and awareness.

Managers' environmental knowledge and values are important factors in enhancing environmental protection. Ashford (1993: 277) mentions that "the key to success in pollution prevention is to influence managerial knowledge of and attitudes toward both technological change and environmental concern" (cited in Fryxell & Lo, 2003). Many previous research studies base the "voluntary" disclosure of environmental information of corporations on legitimacy and stakeholder theory (see, for example, Boesso & Kumar, 2007; Jenkins & Yakovleva, 2006; Liu & Anbumozhi, 2009). However, if environmental information is disclosed as a result of external pressure, such disclosures are more "solicited" than "voluntary" (Van der Laan, 2009). Since the definition of "voluntary" implies actions that are done by "one's own free will" (Voluntary, 2013), "voluntary" disclosures should therefore stem from internal personal values and knowledge that shape management's perception of the magnitude of the environmental problem and their responsibility towards it.

Environmental knowledge is defined by Fyxell and Lo (2003: 48) as "a general knowledge of facts, concepts, and relationships concerning the natural environment and its major ecosystems," while values are "durable concepts or convictions which relate to the desired behaviour, unfold in various situations, provide orientation when evaluating events and are organised in an order of relative importance" (Zso'ka, 2006: 323). The value-belief-norm (VBN) theory of environmentalism suggests that "personal moral norms are the main basis for individuals' general predispositions to pro-environmental action" (Stern, 2000: 413).

One of the main factors influencing managerial knowledge and values is, in fact, education. According to Seoudi and El-Bassiouny (2010: 4):

> Concern over ethical standards in business has been on the rise for several decades now, and more so in the last decade, after the colossal failings associated with the ethical downfall of Enron, Worldcom and others. There seems to be agreement that the problem of low ethical standards in business is systematic rather than isolated incidents resulting from individual weakness. Scholars and thinkers have reflected on the causes of ethically questionable conduct in business. Some studies have proposed that the causes lie in the general level of ethical standards in society while others find that business education is the culprit. Many have voiced their belief that business schools have an important role to play not only in causing but also in preventing further future ethical shortfalls in business. As such, many business schools have added ethics instruction to their curricula, which has in turn stimulated a significant stream of research on the effectiveness of ethics instruction in both the management and pedagogy literature.

There is growing global concern for CSR and ethics education. Although the nascent literature on the topic is growing in the developed Anglo-Saxon world, similar growth is not documented in the developing world. This could be coupled with the complexity of research findings. For instance, some studies found that there is no correlation between students' prior knowledge of business ethics and students' awareness of ethical issues (e.g. Tormo-Carbó et al., 2016). The results could, of course, be sample-specific. Also, the results could be due to the methodologies used in teaching in this particular case. However, in all cases, this raises questions and concerns as to which methodologies and pedagogies should be used in ethics, corporate social responsibility and sustainability (ECSRS) education.

Recent literature has underscored the importance of looking into the contextualization of CSR in developing countries, and especially in the African continent, and found that CSR reporting in Sub-Saharan Africa had been primarily philanthropic (Kuhn et al., 2015).

Mirroring the diversity in CSR definitions (Sarkar & Searcy, 2016), the literature on CSR education has also been diverse. Sarkar and Searcy (2016: 1423) note that there is lack of a consistent "normative basis" for CSR, which resulted in "definitional heterogeneity" as related to the concept. The dimensions of CSR were identified in the study as "economic, social, ethical, stakeholders, sustainability and voluntary" (Sarkar & Searcy, 2016: 1423). It seems plausible to assume, therefore, that the educational basis for ECSRS would also be heterogeneous.

Business education, particularly postgraduate programs such as MBA programs, have been moving toward internationalization, the extensive use of scenario-based learning and case studies, and focusing on the development of students' character and skills (Segon & Booth, 2009). The question remains, however, with these trends, what elements of ECSRS education are necessary to develop students' ethical awareness and propensity to always choose to do the "right thing"? Education is a service, and the output is graduates with certain skills bases. How can educators foster moral character in their students? Can they teach their students particular ethical decision-making techniques? Ethics education and the Giving Voice to Values (GVV) pedagogy start with this premise – that ethics can be taught. Stubbs and Schapper (2011: 260) "argue that any efforts that challenge the accepted 'business as usual' model of business education can make a difference no matter how small."

Studies, on the international level, have used different methods of analysis. For instance, some studies have compared the presence of ECSRS topics in the textbooks of a particular discipline (e.g. Gordon, 2011). Some other studies have attempted to identify thematic approaches to CSR education in particular continents, e.g., Europe (see e.g., Matten & Moon, 2004) or particular countries, e.g., Portugal (see e.g., Branco & Delgado, 2016). With the emergence of CSR in developing countries as an essential field of study (Jamali & Karam, 2016), the study of CSR education in the MENA region, therefore, is both timely and warranted.

This book presents a portfolio of applied case studies in ECSRS education for different countries in the Middle East and North Africa (MENA) region. These case studies can serve as best practice examples for the multiplicity of institutions in the region as well as globally. Country contributions represented in this publication come from Egypt, Morocco, Qatar, and the Sultanate of Oman.

In Chapter 1, which is a contribution of Sultan Qaboos University in Oman, the authors highlight a citation-based systematic literature review of ECSRS education in the MENA region. They conclude that there is need for insightful qualitative research into ECSRS in the varied geo-political spheres of the region. They also highlight that the nascent literature needs much further research.

Chapter 2 presents an overview of CSR in the developing world and sheds light on the Egyptian context for CSR. The chapter then outlines the implications that the research results have for ECSRS education in the MENA region.

Chapter 3 presents a case on the implementation of the GVV methodology in a business ethics course in Egypt as an example to similar contexts in the MENA region.

Chapter 4 contributes to the book by presenting a detailed highlight of education for sustainable development and its contribution to higher education institutions and their utilized pedagogies.

Chapter 5 presents the case of ECSRS education in Morocco as a unique context in the MENA region. The chapter concludes that the implementation of business ethics in Moroccan higher education, as a sample from the MENA region, is still limited.

Finally, Chapter 6 presents the implementation of ECSRS in the College of Business and Economics at Qatar University.

Despite the chapter contributions coming from the same region, it is safe to say that the MENA region is far from being monolithic. Varied contextual factors define how ECSRS is perceived in different contexts. For example, while the connotations of CSR in Egypt mirror an immature development to date, a different reflection is seen in the Moroccan case.

We hope our book serves as a starting point for stirring discussions around how ECSRS education can be taught, especially with reference to the Giving Voice to Values (GVV) methodology and pedagogy.

Noha El-Bassiouny, Dina El-Bassiouny,
Ehab K. A. Mohamed, Mohamed A. K. Basuony,
Egypt, January 2020

References

Aerts, W. & Cormier, D. (2009). Media legitimacy and corporate environmental communication. *Accounting, Organizations and Society, 34*, pp. 1–27.

Al-Abdin, A., Roy, T., & Nicholson, J. (2018). Researching corporate social responsibility in the Middle East: The current state and future directions. *Corporate Social Responsibility and Environmental Management, 25*, pp. 47–65.

Arce, D., & Gentile, M. (2015). Giving Voice to Values as a leverage point in business ethics education. *Journal of Business Ethics, 131* (3), pp. 535–542.

Ashford, N. A. (1993). Understanding technological responses of industrial firms to environmental problems: Implication for government policy, in J. Schot & K. Fischer (eds.), *Environmental Strategies for Industry: International perspectives on research needs and policy implications* (Island Press, Washington, D.C.), pp. 277–310.

Barbu, E. M., Dumontier, P., Feleagă, N., & Feleagă, L. (2011). Mandatory environmental disclosures by companies complying with IAS/IFRS: The case of France, Germany and the UK. Cahier de recherche du CERAG 2011–09 E2. Available online at https://halshs.archives-ouvertes.fr/halshs-00658734/document. Last accessed 1 April 2018.

Boesso, G., & Kumar, K. (2007). Drivers of corporate voluntary disclosure: A framework and empirical evidence from Italy and the United States. *Accounting, Auditing & Accountability Journal, 20* (2), pp. 269–296.

Branco, M., & Delgado, C. (2016). Corporate social responsibility education and research in Portuguese business schools, in D. Turker et al. (eds.) *Social Responsibility Education across Europe: CSR, sustainability, ethics, and governance*, doi: 10.1007/978-3-319-26716-6_10.

Campbell, D. (2004). A longitudinal and cross-sectional analysis of environmental disclosure in UK companies—a research note. *The British Accounting Review, 36*, pp. 107–117.

Cheah E., Jamali, D., Johnson, J., & Sung, M. (2011). Drivers of corporate social responsibility attitudes: The demography of socially responsible investors. *British Journal of Management, 22* (2), pp. 305–323.

Cho, C., Patten, D. M., & Roberts, R. W. (2006). Corporate political strategy: An examination of the relation between political expenditures, environmental performance, and environmental disclosure. *Journal of Business Ethics, 67*, pp. 139–154.

Clarkson, P. M., Richardson, G. D., & Vasvari, F. P. (2007). Revisiting the relation between environmental performance and environmental disclosure: An empirical analysis. *Accounting, Organizations and Society*, doi:10.1016/j.aos.2007.05.003.

Fryxell, G., & lo, C. (2003). The influence of environmental knowledge and values on managerial behaviours on behalf of the environment: An empirical examination of managers in China. *Journal of Business Ethics*, *46*, No. 1, pp. 45–69.

Gentile, M. (2012). Values-driven leadership development: Where we have been and where we could go. *Organization Management Journal*, *9* (3), pp. 188–196.

Gordon, I. (2011). Lessons to be learned: An examination of Canadian and U.S. financial accounting and auditing textbooks for ethics/governance coverage. *Journal of Business Ethics*, *101* (1), pp. 29–47.

Gray, R., Javad, M., Power, D. M., & Sinclair, C. D. (2001). Social and environmental disclosure and corporate characteristics: A research note and extension. *Journal of Business Finance & Accounting*, *28*, No. 3 & 4, pp. 327–356.

Jamali D., & Neville, B. (2011). Convergence versus divergence of CSR in developing countries: An embedded multi-layered institutional lens. *Journal of Business Ethics*, *102* (4), pp. 599–621.

Jamali, D., & Karam, C. (2016). Corporate social responsibility in developing countries as an emerging field of study. *International Journal of Management Reviews*, *00*, pp. 1–30.

Jamali, D., & Sidani, Y. (2012). Introduction, in Jamali, D. and Sidani, Y. (eds.) *CSR in the Middle East: Fresh perspectives* (Palgrave Macmillan, London).

Jamali, D., & Sidani, Y. (2013). Does religiosity determine affinities to CSR? *Journal of Management. Spirituality & Religion*, *10* (4), pp. 309–323.

Jenkins, H., & Yakovleva, N. (2006). Corporate social responsibility in the mining industry: Exploring trends in social and environmental disclosure. *Journal of Cleaner Production*, *14*, pp. 271–284.

Jorgensen, B. N., & Soderstom, N. S. (2006). *Environmental Disclosure Within Legal and Accounting Contexts: An international perspective*. Unpublished manuscript. Available online at www0.gsb.columbia.edu/mygsb/faculty/research/pubfiles/2621/Jorgensen_Soderstrom_20070704.pdf. Last accessed 1 April 2018.

Kuhn, A., Stiglbauer, M., & Fifka, M. (2015). Contents and determinants of corporate social responsibility website reporting in Sub-Saharan Africa: A seven country study. *Business and Society*, pp. 1–14. doi: 10.1177/0007650315614234.

Ling, Q. (2007). *Competitive Strategy, Voluntary Environmental Disclosure Strategy, and Voluntary Environmental Disclosure Quality*. Unpublished doctoral dissertation. Oklahoma State University, USA.

Liu, X., & Anbumozhi, V. (2009). Determinant factors of corporate environmental information disclosure: An empirical study of Chinese listed companies. *Journal of Cleaner Production*, *17*, pp. 593–600.

Magness, V. (2006). Strategic posture, financial performance and environmental disclosure: An empirical test of legitimacy theory. *Accounting, Auditing and Accountability Journal*, *19* (4), pp. 540–563.

Matten, D., & Moon, J. (2004). Corporate social responsibility education in Europe, *Journal of Business Ethics*, *54*, pp. 323–337.

Meek, G., Roberts, C. B., & Gray, S. J. (1995). Factors influencing voluntary annual report disclosures by U.S., U.K. and Continental European multinational corporations. *Journal of International Business Studies*, *26*, No. 3, pp. 555–572.

Sarkar, S., & Searcy, C. (2016). Zeitgeist or chameleon? A quantitative analysis of CSR definitions. *Journal of Cleaner Production*, *135*, pp. 1423–1435.

Segon, M., & Booth, C. (2009). Business Ethics and CSR as Part of MBA Curricula: An Analysis of Student Preference, *International Review of Business Research Papers*, 5 (3), pp. 72–81.

Seoudi, I., & El-Bassiouny, N. (2010). Egyptian business students' perceptions of ethics: The effectiveness of a formal course in business ethics. *Journal of Business Leadership*, pp. 29–49.

Stern, P. C. (2000). Toward a coherent theory of environmentally significant behavior. *Journal of Social Issues, 56* (3), pp. 407–424.

Stubbs, W., & Schapper, J. (2011). Two approaches to curriculum development for educating for sustainability and CSR. *International Journal of Sustainability in Higher Education, 12* (3), pp. 259–268.

Tormo-Carbó, G., Oltra, V., Seguí-Mas, E., & Klimkiewicz, K. (2016). How effective are business ethics/CSR courses in higher education? *Procedia – Social and Behavioral Sciences, 228*, pp. 567–574.

Van der Laan, S. (2009). The role of theory in explaining motivation for corporate social disclosures: Voluntary disclosures vs 'solicited' disclosures. *Australasian Accounting Business and Finance Journal, 3* (4), pp. 15–29.

Voluntary (2013). In: Oxforddictionaries.com. Available online at www.oxforddictionaries.com/definition/english/voluntary. Last accessed 10 November 2013.

Zso'ka, A. (2006). Consistency and "awareness gaps" in the environmental behaviour of Hungarian companies. *Journal of Cleaner Production, 16*, pp. 322–329.

1 A citation-based systematic literature review on ECSRS education in MENA region countries

Nisar Ahmad, Helena Knight and Irfan Butt

Introduction

The last couple of decades have seen a renewed interest from both the academic and the practitioner community in the role of ethics in business education, as a potential institutional antecedent of ethical business practice. The onset of the new millennium marked by globalization, and pervasive and increasingly complex social problems often attributable to dubious business practices, resolutely marked the end of the ambivalence of the role of business schools to promulgate ethical business practices (Smith, 2003). Early critics called out European and North American business schools as co-responsible for the aftermath of the infamous corporate scandals that uncovered unprecedented levels of unethical business practices (Adler, 2002; Ghoshal, 2005). Business schools were accused of having morphed into "brainwashing institutions" preoccupied with shareholder primacy, as prescribed by the Friedmanite ideology (see Friedman (1970) for Friedmanite ideology; see Matten and Moon (2004) for an overview of business schools' criticisms) that perpetuated the amoral management stalemate in the late twentieth-century business practice (Carroll, 2001). Management education at universities was desperately out of sync with the emergent values-driven society that demands business responsibilities to transcend beyond those related to shareholders (Doh & Guay, 2006).

The landscape began to change significantly following the introduction of the Principles for Responsible Management Education (PRME) by the United Nations (UN) in 2002, which endorsed education as one of the underpinning principles of sustainable development. The advocacy by one of the key global institutional players, which stated that academia represents a key institution that "most directly act as drivers of business behavior" (United Nations Compact, 2007) attracted the attention of a variety of actors. Universities from across the globe became PRME signatories (UNPRME, 2019), pledging to integrate the different principles to their operations. An increasing number of scholars began to examine the different factors of ECSRS education. Matten and Moon (2004) provide one of the earliest comprehensive overviews of the state of business ethics and responsibility education in Europe. The authors reject the view of the lack of capacity in European business schools to develop CSR education. CSR course provision is common and takes the form of either individual modules or specialized programs, typically at the executive and postgraduate levels. "Mainstreaming" CSR education, i.e., embedding it in the core of business education for optimal outcomes in the future of business, happens mainly through offering optional modules, but also embedding CSR into other modules and courses, and offering special seminars, industry speakers, and conferences. Practitioner speakers emerged as the most popular teaching method and faculty members as the focal driver of CSR education in European universities.

In the decade since the Matten and Moon (2004) review, and with the institutional support established, research on ethics in management education began to grow steadily. Consistent with Matten and Moon (2004), scholars have considered the different aspects related to institutional matters (Maloni et al., 2012; Wright & Wilton, 2012; Amaral et al., 2015) and the curriculum side (Niu et al., 2010; Nicholls et al., 2013; Lozano et al., 2015) of ethics integration in universities. Matters of terminology have also been considered. Matten and Moon (2004) suggest that in addition to the prevalent CSR terminology, ethics and sustainability are commonly employed. A study of the top 50 business schools listed by the *Financial Times* in 2006 also found that the most frequent combination in study requirements by MBA students is ethics, CSR and sustainability, or ECSRS, as the topic will be referred to in the rest of the chapter (Christensen et al., 2007).

Much of the existing studies focus on a single level of analysis, while the lack of multilevel analysis has been identified as a key gap in CSR research in general (Aguilera et al., 2007; Aguinis & Glavas, 2012). In the ECSRS education field, Setó-Pamies and Papaoikonomou (2016) integrate the institutional level (faculty, university), the curricular level (course design, modules) and the instrumental level (specific methodologies) as a multiprong approach to improve the learning environment, and so instill prosocial attitudes, knowledge and behaviors in business school graduates. However, the framework omits an individual level of analysis, despite the incessant calls in extant literature for greater integration and improved understanding of the individual level as a single unit of analysis and as part of a multilevel study (Aguinis & Glavas, 2012). A more detailed understanding is crucial as it is the beliefs and attitudes of individuals that are the key predictors of outcomes from interventions (Ajzen & Fishbein, 1975; Ryan & Deci, 2000). Moreover, the emerging concept and methodology of Giving Voice to Values suggests that it is the individual level, the students' personal beliefs and attitudes, and their education and training on how to enact their values effectively, that essentially determines whether the aggregate of macro- and meso-level measures results in a sustainable business practice.

Yet, as lamented by Painter-Moorland and Siegers (2018: 807), "What is often missed in our considerations of ethics teaching is what our students already value when they walk into our classrooms." Research thus needs to integrate students' most basic beliefs and guide them to critically reflect upon them, to ensure that they align with the institutional and organizational-level efforts for better future business.

Moreover, as is the case with the general international CSR literature (Aguinis & Glavas, 2012; Pisani et al., 2017), much of the developments and research within the ECSRS education discipline have happened in Western countries. None of the CSR literature reviews focused on the region provide any insights into the state of ECSRS education nor do they highlight education as an area of concern for driving ethical business practice in the MENA countries. This appears to support the declaration by Izraeli made over two decades ago (1997: 1557) that "business ethics education is not institutionalised in academia in the Middle East" and "business ethics courses are not taught, very little research being conducted, there are not any specific publications on the issues, nor any regular training on business ethics." El-Bassiouny et al. (2018) analyzed the status of integrating education for sustainable development in the Arab region for management education. To the best of our knowledge, there is no systematic literature review on the topic of ECSRS education in the Middle East and North Africa (MENA).

A systematic literature review (SLR) is a well-defined process with a comprehensive search technique, research questions, data extractions, and data presentations (Kitchenham et al., 2009) that is invaluable to provide a general picture of the developments in a particular field of study. The main advantage of an SLR would be that it is objective in article selection, all-inclusive, exhaustive, and repeatable, to name a few. Further, it has an explicit criterion for searching papers, including a thorough evaluation and discussion on quality research on a particular topic. SLRs were initially carried out mostly in medical science. However, it has now become a standard in many disciplines. Several studies using this methodology have been published in top-tier journals across many disciplines (e.g., Calma & Davies, 2016; Liu et al., 2013; Liket & Simaens, 2015) with a comprehensive review on CSR published in the *Journal of Management* by Aguinis and Glavas (2012).

Keeping in view the above discussion and to address the lacuna in ECSRS literature, the primary objective of this chapter is to carry out a systematic literature review based on citation analysis and a basic content analysis focused on ECSRS education in the MENA region. Accordingly, the study provides an initial outline of the developments in the field to date using a bibliometric lens. We examined 267 peer-reviewed articles which are focused on CSR in the MENA region. The study provides the leading sources of knowledge in the forms of the most influential papers, authors, and journals. The bibliometric data is complemented by a content analysis of papers that address education, as a specific subset of the overall ECSRS sample. Vital insights are provided into ECSRS education literature in the MENA region through an outline of prevalent research orientations, research design, data collection method, country, ECSRS concept, and level of analysis.

Accordingly, the study makes the following value-added contributions. A comprehensive, multilevel framework for analyzing ECSRS education is offered. This is achieved by integrating an individual level of analysis to the framework proposed by Setó-Pamies and Papaoikonomou (2016). This addition is supported through the results of the content analysis, which show that the individual level in terms of student beliefs and attitudes is the most widely addressed topic in the sample articles. This is the case even though (a) the individual level is the least analyzed level in CSR research (Aguinis & Glavas, 2012), and (b) the recently developed multilevel framework for the integration of ECSRS in management education omits the individual level (Setó-Pamies & Papaoikonomou, 2016). Accordingly, the chapter contributes insights towards the bridging the divide of the micro (individual) and macro (institutional) levels in management studies (Aguilera et al., 2007; Aguinis & Glavas, 2012). We further demonstrate that sustainability as a concept underpins the majority of the education-related studies reviewed in the region. In terms of geopolitical context, ECSRS education research focuses primarily on issues related to Saudi Arabia, UAE, Egypt, and Lebanon, with no studies found for Algeria, Bahrain, Iraq, Jordan, Kuwait, Libya, Morocco, Syria, or Tunisia.

The rest of the study is organized as follows: section 2 provides the methodology used in the study. Citation analysis is discussed in section 3. Section 4 provides the detailed content analysis

Methodology

The primary objective of the current chapter is to carry out a systematic literature survey on CSR, specifically for education in the MENA countries. The analysis has been carried out using a systematic search process in searching for relevant articles. This is followed by a citation and content analysis of the selected education articles. A brief discussion about these methodologies is described in this section.

Article selection process

The literature related to CSR in the MENA region has been searched through well-known databases, namely ScienceDirect (SD), ProQuest Central (PQC), and Business Source Complete (BSC). The selection has been carried out in multiple stages. In the first stage, title and abstract of the academic articles have been searched for any of the following: "corporate social responsibility," "CSR,"] "corporate citizenship," "business ethics," "corporate responsibility," "ISO 26000," and "sustainability." The search returned a total of 63,624 from PQ, 8,343 from BSC, and 1,463 titles from SD. In the second stage, the articles from the first stage are searched for MENA region countries in the abstract and title containing any of the following words: Algeria, Bahrain, Egypt, Iran, Iraq, Israel, Jordan, Kuwait, Lebanon, Libya, Morocco, Oman, Qatar, Saudi Arabia, Palestine, Syria, Tunisia, "United Arab Emirates," Yemen, MENA, GCC, Dubai, Arab, "Middle East and North Africa," "Middle East," Gulf. The number of articles was considerably reduced to 1,320, with 950 from PQ and 370 from BSC. The articles were then taken into "RefWorks" (reference management online system) for the removal of duplicates. The remaining articles (1,249) were then moved to Microsoft Excel and given a unique identification number (ID) for further analysis.

In the next stage, the articles (title and abstract) were manually screened by two independent researchers for relevancy. Multiple papers were found to relate to the macroeconomic impact of environment and sustainability of a particular industry on the economy. Those articles were removed as irrelevant to the scope of our project. For instance, "Renewable and non-renewable energy consumption, environmental degradation and economic growth in Tunisia" survived our search criterion since it contains our search words in the titles, but it is out of the scope of this chapter. Similarly, there were several papers from the medical sciences related to the sustainability of particular medicines. For example, the paper "Review: Animal health and sustainable global livestock systems" is clearly out of the scope of this chapter and was therefore removed from the sample. Further, to control for quality, the articles that do not appear in Web of Science (ISI), Australian Business Deans Council (ABDC), or Scopus have been removed. After careful cross-checks of the articles by the two researchers,

Table 1.1 Title and abstract search results

Searching	*PQ*	*BSC*	*Total*
Stage 1: Abstract and title search for corporate social responsibility words	63,624	8,343	71,967
Stage 2: Title and abstract search for MENA region countries	950	370	1,320
Stage 3: Filtering: duplicates removed, academic journal, peer-reviewed, English language	901	348	1,249
Stage 4: Relevant articles after manually skimming articles for relevant articles (done independently by two researchers on the field)	131	136	267
Articles for citation analysis	267		
A subset of ECSRS education articles for content analysis	23		

267 articles survived quality and the relevancy check. To carry out the content analysis of ECSRS education, a subset of the sampled articles was selected. The articles were identified by carefully searching the titles and abstracts of the main dataset.

Citation analysis

Citation analysis is a way of identifying the relative importance of article/ authors/journals by calculating the number of times that peers have cited an article. The emergence of online academic databases on the Internet has made it possible to perform citation analysis effectively. Google Scholar has become an attractive source of citation numbers since its launch in 2004. Falagas et al. (2008) provide a comprehensive comparison of the content coverage and practical utility of PubMed, Scopus, Web of Science, and Google Scholar. Walters (2009) presents a comparative performance of Google Scholar and 11 other bibliographic databases. They report that in terms of both recall and precision, Google Scholar is superior to the subscription databases. Since these findings are based on a sound evaluation process, in this review we adopt the Google Scholar citation database. Each article is searched by Google Scholar, which immediately gives a citation count.[1] This appears to be better than the Web of Science since it not only includes the citations from the final published version of the article but also adds the citations from the working paper of the same article (Harzing & Alakangas, 2016; Harzing & Van Der Wal, 2009).

Content analysis

This stage is a quantitative approach towards producing an informative source of knowledge from an open-ended data of each paper that are well explored, categorized, and evaluated. "Content analysis is a method for analysing the content of a variety of data, such as visual and verbal data. It enables the reduction of phenomena or events into defined categories to better analyse and interpret them" (Harwood & Garry, 2003). In recent years, researchers have used content analysis for SLR to identify trends of publication in a discipline. Content analysis will be used to read, classify, and summarize the selected papers to extract the most relevant information and produce informative data or measures. This will clarify the most critical aspects of the status quo in the literature by quantifiable measures and visualized information leading to the identification of research gaps.

Results of citation analysis

This section presents the results of the citation analysis with a specific focus on publication trends, the most influential journals, authors, and articles.

Trends in publications and citations

Izraeli (1997) is the seminal paper that examined business ethics in the MENA region. The article, titled "Business ethics in the Middle East," is published in the *Journal of Business Ethics* (JBE) and discusses different concepts of business ethics in the Middle East, which includes the stereotypes about business ethics in the region. Aside from claiming that business ethics is not institutionalized in academia in the region, the author talks about cultural issues in business ethics, and provides some trends that will impact business ethics in the region.

Table 1.2 provides the time frame of articles and citations. There was a steady growth in the publications on ECSRS in the MENA region until 2008, which accounts for 12% of the total articles in our dataset. This is an approximation that shows about 6% of the total number of articles in our database have been published during the period 1997–2003 and another 6% were published between 2004 and 2008. The number increased to 27% in the next time frame (2009–13). In the last four years, a tremendous growth in the literature has occurred, and approximately 61% of articles were published during this time period. This corresponds with the increase in interest in ECSRS overall in the MENA region reported in the literature (Al-Abdin et al., 2018). In the period between 2004 and 2008, only 6% of all articles were published, yet these articles have 18.1 average citations per year, which represents 37% of the overall dataset. This is in stark contrast to the next period (2009–2013), which saw 27% of all articles published, yet these studies have only been cited on average 5.4 times per year. The significantly higher number of average citations in this period is attributed to the publication of the most cited paper in our sample, titled "Corporate social responsibility (CSR): Theory and practice in a developing country context" in the JBE (Jamali & Mirshak, 2007).

The most influential journals publishing CSR research in the MENA region

The selected articles in the analysis have been published in 163 different journals. Figure 1.1 shows the ranking of the top 10 journals based on total

Table 1.2 Time frame of publications and citations

Time frame	*Articles (no.)*	*Articles (%)*	*Citations (no.)*	*Citations (%)*	*Average citations per year*
1997–2003	16	6	1,664	17	5.8
2004–08	16	6	3,570	37	18.1
2009–13	73	27	3,192	33	5.4
2014–18	162	61	1,316	14	2.9
Total	267	100	9,742	100	4.7

citations accumulated in ECSRS in MENA. The JBE is ranked number one by collecting 4,002 citations, which accounts for 41% of the total citations on the topic. Out of 267 articles, 27 articles have been published in this journal. Recently, the *JBE* has been included in the *Financial Times* ranking of the top 50 journals (FT50). It is an encouraging indication that papers focusing on MENA regions are being published in well-reputed journals. The second in the list is *Corporate Governance: An International Review*. Only one article has been published in the journal, but it accumulated 6% of the total citations. The title of the article is "Corporate governance and corporate social responsibility: Synergies and interrelationships" – it talks about the interrelationships between corporate governance and corporate social responsibility by reviewing theoretical literature and by investigating the relationship using sample firms operating in Lebanon. The *Social Responsibility Journal* (SRJ) is ranked third in the list in terms of some publications and citations. Eighteen articles have been published in this journal, and they only accumulated 5% total citations. This is because the SRJ is not listed in Web of Science and it is ranked B in the ABDC list of journals.

Figures 1.2a, 1.2b, and 1.2c show the distribution of papers by impact factor, ABDC, and Scopus indexed journals, respectively. About 64% of the articles are not listed in Web of Science, and about 7% are published in journals with an impact factor equal to or more than 5. According to ABDC ranking, less than 1% of articles are published in A★ journals, whereas 45% of articles are not listed in this ranking. According to the Scopus indexing, approximately 71% of the articles are published in the

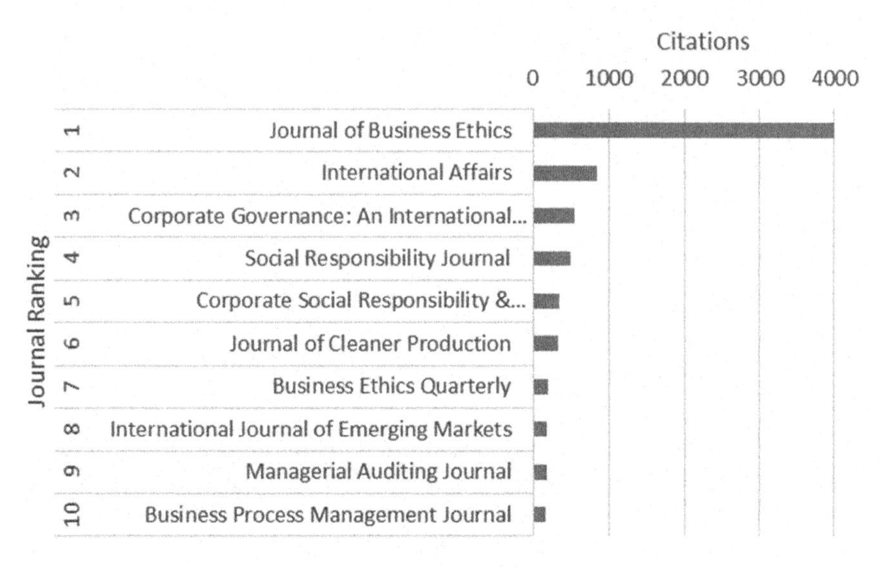

Figure 1.1 Top 10 most influential journals based on total citations on ECSRS in the MENA region.

2 a: Distribution of Articles in ABDC list of Journals

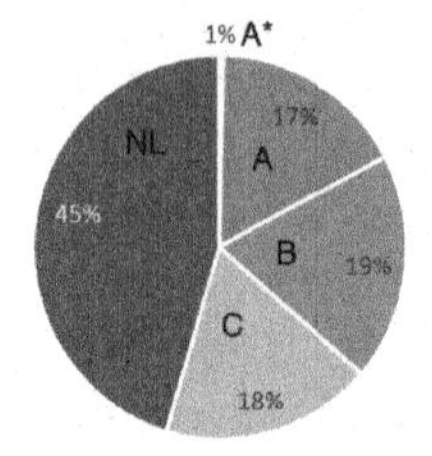

2 b: Distribution of Articles by Impact Factor

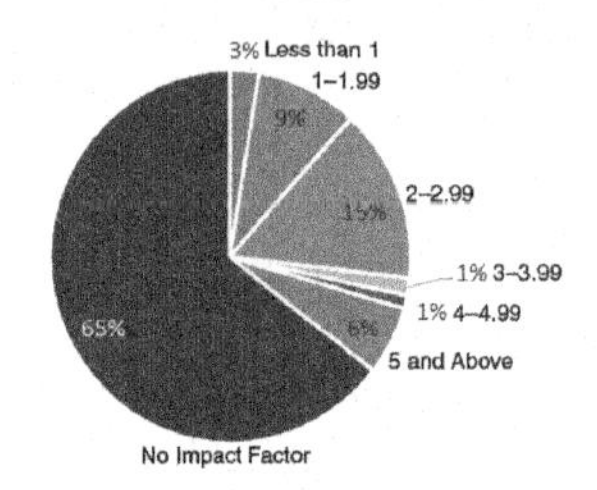

2 c: Distribution in Scopus Index Journals

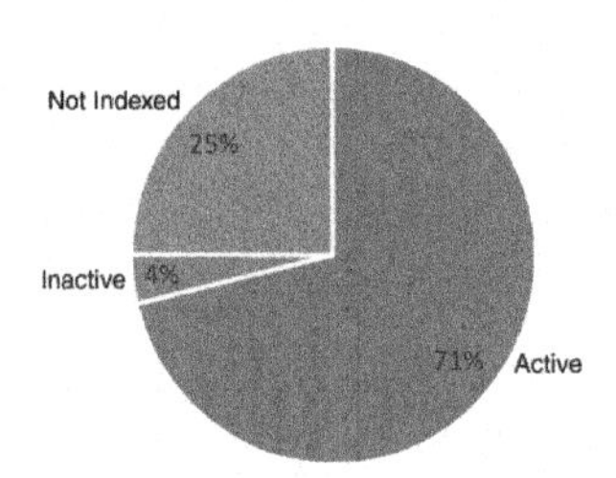

Figure 1.2 Distribution of articles based on ABDC, impact factor, and Scopus status.

active journals bracket, whereas 25% of the articles are not listed in Scopus.

Top authors

Figure 1.3 shows the top 20 most prolific authors in the field based on total citations received on the article published on the topic. The size of the bubble indicates the relative number of citations received by individual authors. Dima Jamali is the most cited author, with 2,358 citations. Ramez Mirshak and Jedrzej George Frynas are at number 2 and 3, with 1,023 and 849 citations respectively.

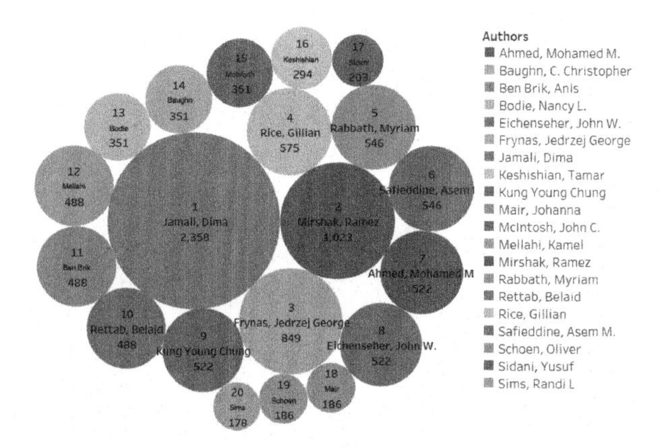

Figure 1.3 Top 20 most prolific authors.

Most influential articles

Table 1.3 provides information about the top 20 influential articles in the field of CSR in the MENA region. As a general trend, it can be seen that the papers published earlier accumulate more citations in total compared with the papers published later. Therefore, for meaningful comparison between old and new articles, it is important to rank the papers according to average citation per year, as shown in column 1. The most influential paper is titled "Corporate social responsibility (CSR): Theory and practice in a developing country context" authored by Dima Jamali and Ramez Mirshak, published in the JBE. The paper used two robust CSR conceptualizations based on Carroll (1979) and Wood (1991). The study critically examines the CSR approach and philosophy of eight companies that were active in CSR in Lebanon at the time. The study finds a lack of a systematic, focused, and institutionalized approach to CSR. CSR practice in Lebanon is still grounded in the context of philanthropic action. Our findings further show that 10 of the top 20 papers are published in the JBE, showing the recognition given to the topic by a very highly rated journal.

Results of the content analysis of ECSRS education articles in the MENA region

In this section, a detailed content analysis of papers related to ECSRS education in the MENA region is provided to identify any gaps in the literature. We found 23 out of 267 papers that examine some aspect of ECSRS education. The studies are categorized based on research orientation, respondents, level of analysis, concept of CSR used, and country of study.

Table 1.3 Top 20 most influential articles

Rank	*ACY*	*TC*	*Title*	*ABDC*	*IF*	*Scopus*	*Reference*
1	85.3	1,023	Corporate social responsibility (CSR): Theory and practice in a developing country context.	A	2.35	Active	Jamali and Mirshak (2007)
2	49.6	546	Corporate governance and corporate social responsibility synergies and interrelationships	A	3.57	NL	Jamali et al. (2008)
3	37.0	37	The impact of corporate social responsibility disclosure on financial performance: Evidence from the GCC Islamic banking sector	A	2.35	Active	Platonova et al. (2018)
4	34.8	348	A study of management perceptions of the impact of corporate social responsibility on organisational performance in emerging economies: The case of Dubai	A	2.35	Active	Rettab et al. (2009)
5	29.4	294	Uneasy alliances: Lessons learned from partnerships between businesses and NGOs in the context of CSR	A	2.35	Active	Jamali & Keshishian (2009)
6	29.3	351	Corporate social and environmental responsibility in Asian countries and other geographical regions	C	2.85	Active	Baughn et al. (2007)
7	28.8	575	Islamic ethics and the implications for business	A	2.35	Active	Rice (1999)
8	24.6	221	The CSR of MNC subsidiaries in developing countries: Global, local, substantive or diluted?	A	2.35	Active	Jamali (2010)
9	24.4	122	Corporate social responsibility and financial performance in Saudi Arabia	B	NL	Active	Mallin et al. (2014)
10	18.0	162	Sustainable supply chains: A study of interaction among the enablers	B	NL	Active	Nishat Faisal (2010)
11	18.0	54	Reduction of food waste generation in the hospitality industry	NL	5.71	Active	Pirani & Arafat (2016)

Rank	*ACY*	*TC*	*Title*	*ABDC*	*IF*	*Scopus*	*Reference*
12	17.5	140	Market orientation, corporate social responsibility, and business performance	A	2.35	Active	Brik et al. (2011)
13	16.3	261	Business students' perception of ethics and moral judgment: A cross-cultural study	A	2.35	Active	Ahmed et al. (2003)
15	15.5	186	Successful social entrepreneurial business models in the context of developing economies	C	NL	Active	Mair & Schoen (2007)
16	15.5	124	Sustainable entrepreneurship: Is entrepreneurial will enough? A north-south comparison	A	2.35	Active	Spence et al. (2011)
17	14.3	43	Competing through employee engagement: A proposed framework	B	NL	Active	Al Mehrzi & Singh (2016)
18	13.3	80	Co-creating sustainability: Cross-sector university collaborations for driving sustainable urban transformations	NL	5.71	Active	Trencher et al. (2013)
19	13.0	78	Gendering CSR in the Arab Middle East: An institutional perspective	A	NL	Active	Karam & Jamali (2013)
20	12.3	37	Corporate governance and corporate social responsibility disclosure: Evidence from Saudi Arabia	B	NL	Active	Habbash (2016)

Notes
CY: average citation per year; TC: total citations; ABDC: Australian Business Deans Council Ranking; IF: impact factor from Web of Science; NL: not listed.

Research orientation and design

Figure 1.4 shows the distribution of the identified articles based on research orientation and design. Approximately 70% of the studies are quantitative in nature, followed by 13% of mixed-method studies. Sixty-one per cent of the identified articles use a survey design, whereas 18% studies utilized a combination of a survey and interviews. Just 4% of studies were qualitative. Such a keen lack of qualitative research in the sample studies indicates that this research approach is greatly undervalued. Liket and Simaens (2015) suggest that such a state of affairs may be problematic as qualitative research is crucial for theory development. Table A1 in the appendix provides detailed content of all 23 studies.

Level of analysis and CSR concept

As previously outlined, the majority of existing studies on CSR overall, and on ECSRS education, focus on a single level of analysis. Setó-Pamies and Papaoikonomou (2016) integrate the institutional level (faculty, university), the curricular level (course design, modules) and the instrumental level (specific methodologies) into a multi-pronged approach to improve the learning environment, and so instill pro-social attitudes, knowledge, and behaviors in business school graduates. These levels are integrated to provide mutually reinforcing levels through which to successfully integrate ECSRS education. In this respect, CSR-oriented culture (instrumental level) needs to exist to

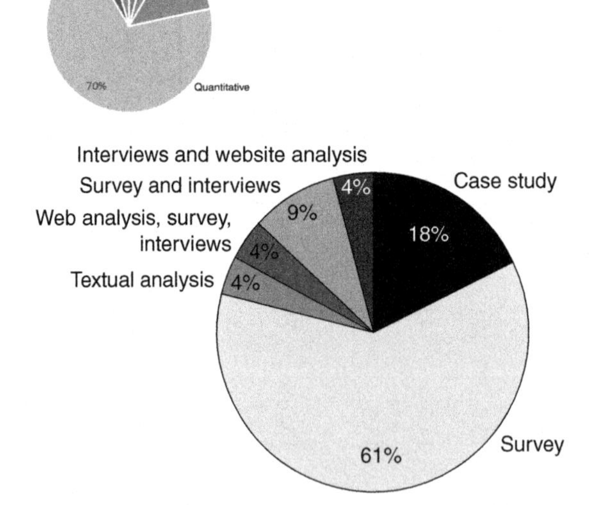

Figure 1.4 Distribution of studies based on research orientation and design.

support the implementation and maintenance of the academic curriculum (curricular level), and the effectiveness of the curriculum will be directly influenced by the specific teaching and learning approaches adopted (instrumental level).

At the institutional level, ethics must be fully integrated into the institution's strategic and operational plans, leadership responsibilities assigned, and appropriate resources allocated. This is often driven by individual faculty members (Matten & Moon, 2004) but may also be advocated by individual institutional administrators. The potential impact of individuals on the successful implementation of ECSRS on the institutional level signals an avenue for researchers in the discipline to begin the bridging of the micro, meso (organizational), and macro (institutional) levels called for in CSR research (Aguilera et al., 2007). The institutional level further includes decisions as to the type of "label" through which the course will be incorporated (Matten & Moon, 2004; Christensen et al., 2007). The curricular level includes decisions in respect of, for example, whether the courses will be stand-alone or concentrated, and whether to remain within a single discipline or adopt a more holistic multi-disciplinary approach (Annan-Diab & Molinari, 2017). The instrumental or pedagogical level then determines through what forms of delivery in terms of pedagogical tools the subject will be delivered.

The content analysis carried out for the purposes of this study indicates a significant shortage of pedagogical studies on ECSRS. This would indicate that even though ECSRS is on the agendas of universities in the MENA region and the curricular-level decisions have been made, there is a shortage of studies on the specific pedagogical tools used to implement the topic. Figure 1.5 illustrates that only 13% of articles, or four out of 23, address the curricular level compared with around 30% for each of the other levels. Moreover, none of the articles address all four levels. One study reports on three levels: institutional level, curricular level, individual level; and seven articles focus on a combination of two levels of analysis, with the most frequently used combination (four articles) analyzing the institutional level and curricular level.

Moreover, the results of the content analysis show that there is in fact a fourth level that needs to be integrated when developing multilevel frameworks. The individual level, which considers normative motives such as alignment to personal values, and individual's attitude to and concern with issues (Aguinis & Glavas, 2012), plays a crucial role in individual students' adoption of pro-ECSRS attitudes and behaviors. Lamborn et al. (1992) define student motivation as "a general desire or disposition to succeed in academic work and in the more specific tasks of school." Ryan and Deci (2000) classify this as "intrinsic motivation"; it is distinct by a high degree of perceived internal control and the activity is undertaken for its own sake, enjoyment, and interest. Pintrich (2003) states that from the motivational theories standpoint, we also need to consider the factors that energize behavior and direct it. Extrinsic motivation is concerned with what makes students engage in and remain engaged in the education process, i.e., the external factors that bear upon the decision-making process (Ryan & Deci, 2000). Accordingly, an individual's desire to learn (individual level) initiates the process, and extrinsic factors (institutional, curricular, and instrumental levels) provide further and continued drive. Motivation as an antecedent factor at the individual level is important, but when combined with engagement facilitated through the rest of the levels, this is where real learning occurs (Saeed & Zingier, 2012). This would suggest that any study on ECSRS education needs to start by determining the perceptions of the students of the proposed topic, as this may be the make or break for the success of the multilevel approach. As entailed in the Giving Voice to Values concept, uncovering students' values related to justice, honesty, and care, as the underpinning principles of moral reasoning, will enable educators to hold meaningful discussions with students rather than attempting to instill prevalent paradigms that may result in students' lack of engagement or outright rejection (Painter, 1998; Arce & Gentile, 2015).

It appears that there is much to learn about overcoming the micro–macro divide in the ECSRS education field in the MENA region.

Our analysis of the articles further indicates that sustainability has superseded CSR as the most prevalent concept in ECSRS education in the region (Figure 1.5), with 11 articles addressing sustainability as a single concept or a part of a group of concepts. This is followed by business ethics, with eight articles. CSR, with only four articles, is the least addressed concept from the major concepts reported in the literature (Matten & Moon, 2004; Christensen et al., 2007; Setó-Pamies & Papaoikonomou, 2016). One article in our sample specifically refers to "consumer ethics," representing an additional potential term in the MENA region landscape. As Figure 1.5 further illustrates, sustainability is most commonly examined at the institutional and curricular levels of analysis, with each of these levels being addressed in six sustainability-oriented papers.

Respondents

Table 1.4 shows the distribution of respondents in the sampled studies. Twelve studies have used only students as the sole respondents, nine of which are linked to the studies that examine the individual level of analysis. Only four studies have used university administrators as respondents and three focused on faculty. None of them examined those respondents using individual-level analysis. However, extant literature posits that successful integration of ethics into an organizational culture is often driven by senior staff, in our case administrators and/or faculty (Aguilera et al., 2007). The oversight in analyzing administrators and faculty from the individual-level platform shows a gap in the ECSRS education literature in the MENA region and a potential future research avenue. Furthermore, only one article in our sample, Taleb (2014), examines the entire set of respondents, which includes students, faculty, and university administration and business managers.

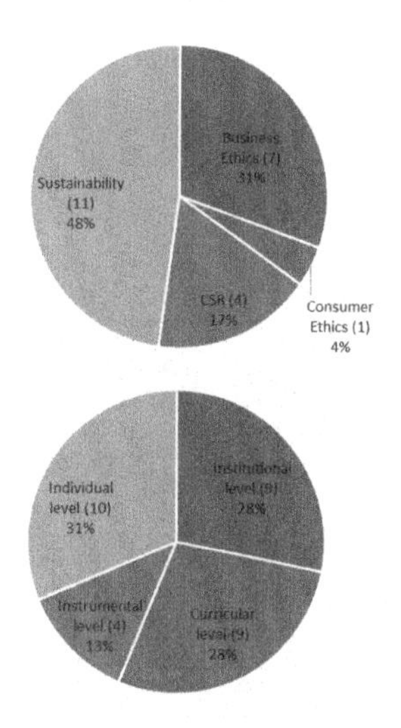

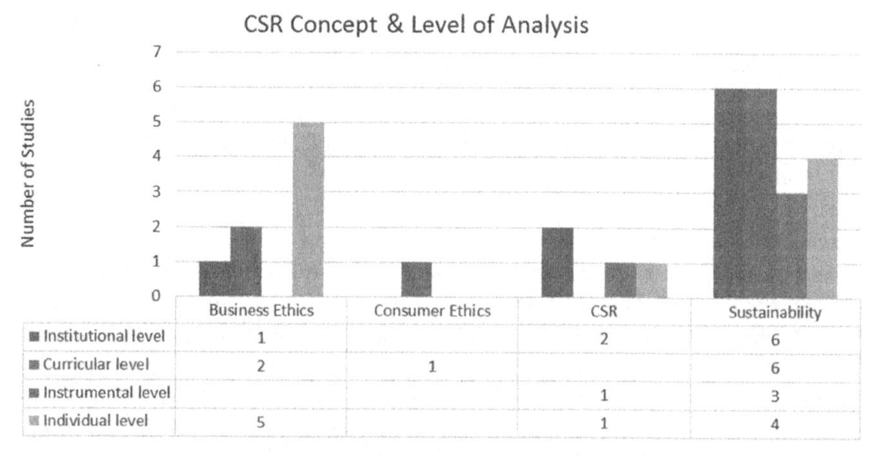

Figure 1.5 CSR concept and level of analysis.

Table 1.4 Respondents in different studies

Respondent	*Number of studies*
Students	12
University administration	4
Faculty and university administration	1
Faculty, business managers	1
Pupils	1
Students, faculty, university administration, business managers	1
Business managers, policymakers	1
Not applicable	2

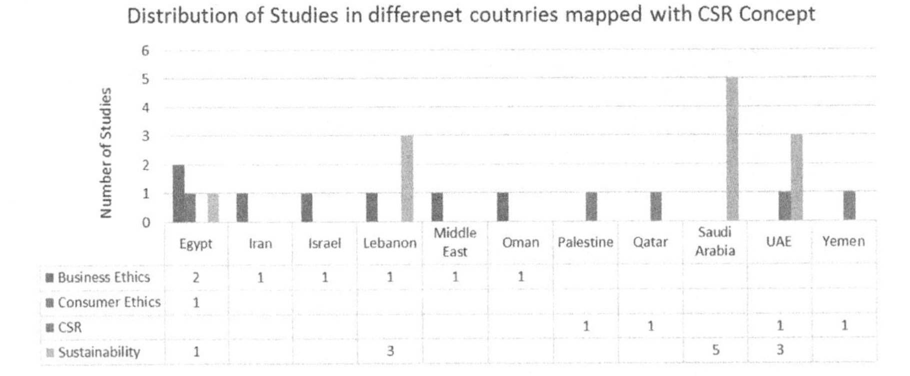

Figure 1.6 Geographical distribution of studies.

Note
No study found for the following countries: Algeria, Bahrain, Iraq, Jordan, Kuwait, Libya, Morocco, Syria, Tunisia.

Country distribution with CSR concept

Previous CSR studies in the Middle East have reported that the majority of ECSRS research originates only in a small handful of countries in the region (Al-Abdin et al., 2018; Jamali & Neville, 2011). The results of our content analysis confirm this state of affairs. As is shown in Figure 1.6, most articles originate from Saudi Arabia (5); Egypt, Lebanon and the United Arab Emirates appear in four articles each. Corresponding with the findings of Al-Abdin et al. (2018), the content analysis shows that no articles originate in Algeria, Bahrain, Iraq, Jordan, Kuwait, Libya, Morocco, Syria, or Tunisia. We can only speculate whether this is due to ECSRS apathy, lack of research output by the relevant institutions in the respective countries, or political upheavals impeding research. Regardless of the reasons, there are considerable research opportunities for scholars to examine the distinct geo-political contexts.

Conclusion and future research avenues

The purpose of this study has been to provide an overview of the extant literature on CSR in the MENA region, and to sketch out the developments in ECSRS education within this region to date. For that purpose, we have carried out an SLR on 267 peer-reviewed articles. In the first step of the SLR process, we carried out a citation analysis to provide a bird's eye view of the CSR literature. The second step consisted of conducting a content analysis of a subset of articles which address ECSRS education in this region.

The study shows that the majority of the extant literature has been published in the 2010s. Especially in the past five years, the number of publications has more than doubled compared with the previous five-year period.

Considering the citation analysis results, the JBE is the most influential journal in the MENA region. Dima Jamali is the most prolific and influential author, having accrued more than double the citations to the next most cited author, Ramez Mirshak. These two authors have jointly produced the most cited article, which was published in the JBE. We found that qualitative research is lacking in ECSRS education studies, showing that the field lacks in-depth insights into the ECSRS education discipline.

The study proposes a comprehensive multilevel framework for analyzing ECSRS education by adding an individual level of analysis to an existing framework developed by Setó-Pamies and Papaoikonomou (2016). The extended framework integrates the individual level with the institutional level, the curricular level, and the instrumental level into a multi-pronged framework to analyze ECSRS education. The individual level in terms of student beliefs and attitudes emerged as the most common level of analysis addressed in ECSRS education research in the MENA region. This is despite the fact that the individual level is the least analyzed level in CSR research (Aguinis & Glavas, 2012), and prior research overviews have neglected this important set of antecedent factors. The study shows that there exists a reasonable degree of insight into students' perceptions; however, the views of other stakeholder are underrepresented.

In addition, sustainability has emerged in the review as the most widely used ECSRS concept in the education field in the MENA region. In terms of geo-political context, ECSRS education research focuses primarily on issues related to Saudi Arabia, UAE, Egypt, and Lebanon, with no study found for Algeria, Bahrain, Iraq, Jordan, Kuwait, Libya, Morocco, Syria, or Tunisia.

In light of the above summary, the scholars in the region need to focus on integrating qualitative research in the research design to provide in-depth insights into the different aspects of ECSRS education in the region. Aside from students' views, the extant literature lacks detailed insights into the perceptions of ECSRS by other key stakeholders, such as faculty, administrators, and the business community. Integrating the diverse voices across the four levels of analysis may provide a useful starting point for bridging the micro- and macro-level gap reported in management literature. Future research can also focus on how and whether integrating ECSRS education in higher education institutions may lead to more tangible positive societal impact.

Finally, our study provides further evidence for Al-Abdin et al.'s (2018) recommendation that there are significant opportunities for future research in the different geo-political contexts in the region that have thus far not produced any study on CSR or on ECSRS education. This is particularly the case for those countries currently experiencing armed conflict and other political disturbances. ECSRS education offers significant opportunities to install the principles of ethical business practice in the future business leaders for when the restructuring of the respective country and re-establishing of its civil society begins.

Table A1 Detailed contents of education-related studies ranked according to average citations per year (ACY)

ID	*Study reference*	*Research orientation*	*Research design*	*Type of data*	*Data collection method*	*Respondent*	*Sample size*	*Data collection country*	*CSR concept*	*Level of analysis*
1	Ahmed et al. (2003)	Quantitative	Survey	Primary	Questionnaire	Students	1,154	Egypt, Finland, Korea, Russia, USA, China	Business ethics	Student personal values
2	Schmidt & Cracau (2017)	Quantitative	Survey	Primary	Questionnaire	Students	265	Qatar, Germany	CSR	Student personal values
3	Abubakar et al. (2016)	Quantitative	Survey	Primary	questionnaire	Students	152	Saudi Arabia	Sustainability	Institutional level, Curricular level, Student personal values
4	Alshuwaikhat et al. (2016)	Quantitative	Survey	Primary	Questionnaire	University administration	Not provided	Saudi Arabia	Sustainability	Institutional level, Curricular level
5	Jouda et al. (2016)	Quantitative	Survey	Primary	Questionnaire	Faculty and university administration	115	Palestine	CSR	Institutional level
6	Marta et al. (2003)	Quantitative	Survey	Primary	Questionnaire	students	360	Egypt, USA	Business ethics	Student personal values
7	Hejase & Tabch (2012)	Quantitative	Case Study	Primary	Questionnaire	Students	262	Lebanon	Business ethics	Curricular level
8	Goby & Nickerson (2012)	Descriptive study	Case Study	NA	NA	Students	Undisclosed fully	UAE	CSR	Instrumental level

ID	Study reference	Research orientation	Research design	Type of data	Data collection method	Respondent	Sample size	Data collection country	CSR concept	Level of analysis
9	Schwartz (2012)	Mixed	Web analysis, survey, interviews	Primary and secondary	Questionnaire, interviews, website, informal discussions	Faculty, Business managers	22 interviews, not available for survey	Israel	Business ethics	Institutional level, Curricular level
10	Rajasekar & Simpson (2014)	Quantitative	Survey	Primary	Questionnaire	Students	378	Oman, India	Business ethics	Student personal values
11	El-Bassiouny et al. (2011)	Mixed	Survey and interviews	Primary	Questionnaire and interviews	Pupils	Not provided	Egypt	Consumer ethics	Curricular level
12	Nejati et al. (2011)	Quantitative	Survey	Primary	Questionnaire	Students	120	Iran	Business ethics	Student personal values
13	Khan et al. (2017)	Comparative study	Case study	Secondary	NA	NA	NA	Saudi Arabia	Sustainability	Institutional level
14	Al-Naqbi & Alshannag (2018)	Quantitative	Survey	Primary	Questionnaire	Students	823	UAE	Sustainability	Student personal values
15	Chuvieco et al. (2018)	Quantitative	Survey	Primary	Questionnaire	Students	1,011	UAE	Sustainability	Instrumental level, Student personal values
16	Koch (2018)	Text analysis	Textual analysis	Secondary	NA	NA	NA	UAE, Qatar, Saudi Arabia	Sustainability	Institutional level, Curricular level
17	Al-Hosaini & Sofian (2015)	Quantitative	Survey	primary	Questionnaire	University administration	136	Yemen	CSR	Institutional level

Continued

Table A1 continued

ID	*Study reference*	*Research orientation*	*Research design*	*Type of data*	*Data collection method*	*Respondent*	*Sample size*	*Data collection country*	*CSR concept*	*Level of analysis*
18	Kanbar (2012)	Quantitative	Survey	Primary	Questionnaire	Students	227	Lebanon	Sustainability	Instrumental level, Student personal values
19	Taleb (2014)	Mixed	Survey and interviews	Primary	Questionnaire and interviews	Students, faculty, university administration, business managers		Saudi Arabia	Sustainability	Curricular level
20	Mezher & Zreik (2000)	Quantitative	Survey	Primary	Questionnaire	University administration	7	Lebanon	Sustainability	Curricular level, Instrumental level
21	Wafik et al. (2011)	Quantitative	Case study	Primary	Questionnaire	Business managers, Policymakers	17	Egypt	Sustainability	Institutional level
22	Almasri & Tahat (2018)	Quantitative	Survey	Primary	Questionnaire	Students	401	Middle East	Business ethics	Student personal values
23	Chidiac El Hajj et al. (2017)	Qualitative	Interviews and website analysis	Primary and secondary	Interviews, observations, website	University administration	Not provided	Lebanon	Sustainability	Institutional level, Curricular level

Note

1 The citation count in this chapter is recorded by the end of 2018.

References

Abubakar, I., Al-Shihri, F., & Ahmed, S. 2016. Students' assessment of campus sustainability at the University of Dammam, Saudi Arabia. *Sustainability*, 8, 59.

Adler, P. S. 2002. Corporate scandals: It's time for reflection in business schools. *Academy of Management Perspectives*, 16, 148–149.

Aguilera, R. V., Rupp, D. E., Williams, C. A., & Ganapathi, J. 2007. Putting the S back in corporate social responsibility: A multilevel theory of social change in organizations. *Academy of Management Review*, 32, 836–863.

Aguinis, H., & Glavas, A. 2012. What we know and don't know about corporate social responsibility: A review and research agenda. *Journal of Management*, 38, 932–968.

Ahmed, M. M., Chung, K. Y., & Eichenseher, J. W. 2003. Business students' perception of ethics and moral judgment: A cross-cultural study. *Journal of Business Ethics*, 43, 89–102.

Ajzen, I., & Fishbein, M. 1975. A Bayesian analysis of attribution processes. *Psychological Bulletin*, 82, 261.

Al-Abdin, A., Roy, T., & Nicholson, J. D. 2018. Researching corporate social responsibility in the Middle East: the current state and future directions. *Corporate Social Responsibility and Environmental Management*, 25, 47–65.

Al-Hosaini, F. F., & Sofian, S. 2015. The influence of corporate social responsibility on customer perspective in private universities. *International Journal of Economics and Financial Issues*, 5, 257–263.

Al-Naqbi, A. K., & Alshannag, Q. 2018. The status of education for sustainable development and sustainability knowledge, attitudes, and behaviors of UAE university students. *International Journal of Sustainability in Higher Education*, 19, 566–588.

Al Mehrzi, N., & Singh, S. K. 2016. Competing through employee engagement: a proposed framework. *International Journal of Productivity and Performance Management*, 65, 831–843.

Almasri, N., & Tahat, L. 2018. Ethics vs IT ethics: a comparative study between the USA and the Middle East. *Journal of Academic Ethics*, 16, 329–358.

Alshuwaikhat, H., Adenle, Y., & Saghir, B. 2016. Sustainability assessment of higher education institutions in Saudi Arabia. *Sustainability*, 8, 750.

Amaral, L. P., Martins, N., & Gouveia, J. B. 2015. Quest for a sustainable university: a review. *International Journal of Sustainability in Higher Education*, 16, 155–172.

Annan-Diab, F., & Molinari, C. 2017. Interdisciplinarity: practical approach to advancing education for sustainability and for the sustainable development goals. *International Journal of Management Education*, 15, 73–83.

Arce, D. G., & Gentile, M. C. 2015. Giving Voice to Values as a leverage point in business ethics education. *Journal of Business Ethics*, 131, 535–542.

Baughn, C. C., Bodie, N. L., & McIntosh, J. C. 2007. Corporate social and environmental responsibility in Asian countries and other geographical regions. *Corporate Social Responsibility and Environmental Management*, 14, 189–205.

Brik, A. B., Rettab, B., & Mellahi, K. 2011. Market orientation, corporate social responsibility, and business performance. *Journal of Business Ethics*, 99, 307–324.

Calma, A., & Davies, M. 2016. Academy of Management Journal, 1958–2014: a citation analysis. *Scientometrics*, 108, 959–975.

Carroll, A. B. 1979. A three-dimensional conceptual model of corporate performance. *Academy of Management Review*, 4, 497–505.

Carroll, A. B. 2001. Models of management morality for the new millennium. *Business Ethics Quarterly*, 11, 365–372.

Chidiac El Hajj, M., Abou Moussa, R., & Chidiac, M. 2017. Environmental sustainability out of the loop in Lebanese universities. *Journal of International Education in Business*, 10, 49–67.

Christensen, L. J., Peirce, E., Hartman, L. P., Hoffman, W. M., & Carrier, J. 2007. Ethics, CSR, and sustainability education in the *Financial Times* top 50 global business schools: baseline data and future research directions. *Journal of Business Ethics*, 73, 347–368.

Chuvieco, E., Burgui-Burgui, M., Da Silva, E. V., Hussein, K., & Alkaabi, K. 2018. Factors affecting environmental sustainability habits of university students: intercomparison analysis in three countries (Spain, Brazil and UAE). *Journal of Cleaner Production*, 198, 1372–1380.

Doh, J. P., & Guay, T. R. 2006. Corporate social responsibility, public policy, and NGO activism in Europe and the United States: an institutional-stakeholder perspective. *Journal of Management Studies*, 43, 47–73.

El-Bassiouny, N., Mohamed, E. K., Basuony, M. A., & Kolkailah, S. 2018. An exploratory study of ethics, CSR, and sustainability in the management education of top universities in the Arab region. *Journal of Business Ethics Education*, 15, 49–74.

El-Bassiouny, N., Taher, A., & Abou-Aish, E. 2011. An empirical assessment of the relationship between character/ethics education and consumer behavior at the tweens segment: the case of Egypt. *Young Consumers*, 12, 159–170.

Falagas, M. E., Pitsouni, E. I., Malietzis, G. A., & Pappas, G. 2008. Comparison of PubMed, Scopus, Web of Science, and Google Scholar: strengths and weaknesses. *The FASEB journal*, 22, 338–342.

Friedman, M. 1970. A Friedman doctrine: the social responsibility of business is to increase its profits. *New York Times Magazine*, 13, 32–33.

Ghoshal, S. 2005. Bad management theories are destroying good management practices. *Academy of Management Learning & Education*, 4, 75–91.

Goby, V. P., & Nickerson, C. 2012. Introducing ethics and corporate social responsibility at undergraduate level in the United Arab Emirates: an experiential exercise on website communication. *Journal of Business Ethics*, 107, 103–109.

Habbash, M. 2016. Corporate governance and corporate social responsibility disclosure: evidence from Saudi Arabia. *Social Responsibility Journal*, 12, 740–754.

Harwood, T. G., & Garry, T. 2003. An overview of content analysis. *The Marketing Review*, 3, 479–498.

Harzing, A.-W., & Alakangas, S. 2016. Google Scholar, Scopus and the Web of Science: a longitudinal and cross-disciplinary comparison. *Scientometrics*, 106, 787–804.

Harzing, A. W., & Van Der Wal, R. 2009. A Google Scholar h-index for journals: an alternative metric to measure journal impact in economics and business. *Journal of the American Society for Information Science and Technology*, 60, 41–46.

Hejase, H. J., & Tabch, H. 2012. Ethics education: an assessment case of the American University of Science and Technology–Lebanon. *International Journal of Islamic and Middle Eastern Finance and Management*, 5, 116–133.

Izraeli, D. 1997. Business ethics in the Middle East. *Journal of Business Ethics*, 16, 1555–1560.

Jamali, D. 2010. The CSR of MNC subsidiaries in developing countries: global, local, substantive or diluted? *Journal of Business Ethics*, 93, 181–200.

Jamali, D., & Keshishian, T. 2009. Uneasy alliances: lessons learned from partnerships between businesses and NGOs in the context of CSR. *Journal of Business Ethics*, 84, 277–295.

Jamali, D., & Mirshak, R. 2007. Corporate social responsibility (CSR): theory and practice in a developing country context. *Journal of Business Ethics*, 72(3), 243–262.

Jamali, D., & Neville, B. 2011. Convergence versus divergence of CSR in developing countries: an embedded multi-layered institutional lens. *Journal of Business Ethics*, 102, 599–621.

Jamali, D., Safieddine, A. M., & Rabbath, M. 2008. Corporate governance and corporate social responsibility synergies and interrelationships. *Corporate Governance: An International Review*, 16, 443–459.

Jouda, A. A., Ahmad, U. N. U., & Dahleez, K. A. 2016. The impact of HRM practices on employees' performance: the case of Islamic University of Gaza (IUG) in Palestine. *International Review of Management and Marketing*, 6.

Kanbar, N. 2012. Can education for sustainable development address challenges in the Arab region? Examining business students' attitudes and competences on education for sustainable development: a case study. *Discourse and Communication for Sustainable Education*, 3, 41–62.

Karam, C. M., & Jamali, D. 2013. Gendering CSR in the Arab Middle East: an institutional perspective. *Business Ethics Quarterly*, 23, 31–68.

Khan, R. A., Al Mesfer, M. K., Khan, A. R., Khan, S., & Van Zutphen, A. 2017. Green examination: integration of technology for sustainability. *Environment, Development and Sustainability*, 19, 339–346.

Kitchenham, B., Brereton, O. P., Budgen, D., Turner, M., Bailey, J., & Linkman, S. 2009. Systematic literature reviews in software engineering–a systematic literature review. *Information and Software Technology*, 51, 7–15.

Koch, N. 2018. Green laboratories: university campuses as sustainability "exemplars" in the Arabian Peninsula. *Society & Natural Resources*, 31, 525–540.

Lamborn, S., Newmann, F., & Wehlage, G. 1992. The significance and sources of student engagement. *Student Engagement and Achievement in American Secondary Schools*, 11–39.

Liket, K., & Simaens, A. 2015. Battling the devolution in the research on corporate philanthropy. *Journal of Business Ethics*, 126, 285–308.

Liu, J. S., Lu, L.Y., Lu, W.-M., & Lin, B. J. 2013. Data envelopment analysis 1978–2010: a citation-based literature survey. *Omega*, 41, 3–15.

Lozano, R., Ceulemans, K., & Seatter, C. S. 2015. Teaching organisational change management for sustainability: designing and delivering a course at the University of Leeds to better prepare future sustainability change agents. *Journal of Cleaner Production*, 106, 205–215.

Mair, J., & Schoen, O. 2007. Successful social entrepreneurial business models in the context of developing economies: an explorative study. *International Journal of Emerging Markets*, 2, 54–68.

Mallin, C., Farag, H., & Ow-Yong, K. 2014. Corporate social responsibility and financial performance in Islamic banks. *Journal of Economic Behavior & Organization*, 103, S21–S38.

Maloni, M. J., Smith, S. D., & Napshin, S. 2012. A methodology for building faculty support for the United Nations Principles for Responsible Management Education. *Journal of Management Education*, 36, 312–336.

Marta, J. K. M., Attia, A., Singhapakdi, A., & Atteya, N. 2003. A comparison of ethical perceptions and moral philosophies of American and Egyptian business students. *Teaching Business Ethics*, 7, 1–20.

Matten, D., & Moon, J. 2004. Corporate social responsibility. *Journal of Business Ethics*, 54, 323–337.

Mezher, T., & Zreik, C. 2000. Current environmental management practices in the Lebanese manufacturing sector. *Eco-Management and Auditing: The Journal of Corporate Environmental Management*, 7, 131–142.

Nejati, M., Amran, A., & Shahbudin, A. S. M. 2011. Attitudes towards business ethics: a cross-cultural comparison of students in Iran and Malaysia. *International Journal of Business Governance and Ethics*, 6, 68.

Nicholls, J., Hair Jr, J. F., Ragland, C. B., & Schimmel, K. E. 2013. Ethics, corporate social responsibility, and sustainability education in AACSB undergraduate and graduate marketing curricula: a benchmark study. *Journal of Marketing Education*, 35, 129–140.

Nishat Faisal, M. 2010. Sustainable supply chains: a study of interaction among the enablers. *Business Process Management Journal*, 16, 508–529.

Niu, D., Jiang, D., & Li, F. 2010. Higher education for sustainable development in China. *International Journal of Sustainability in Higher Education*, 11, 153–162.

Painter, M. 1998. *Collaborative Federalism: Economic reforms in Australia in the 1990s*, Cambridge, Cambridge University Press.

Painter-Morland, M., & Slegers, R. 2018. Strengthening 'Giving Voice to Values' in business schools by reconsidering the 'invisible hand' metaphor. *Journal of Business Ethics*, 4, 807. doi: 10.1007/s10551-017-3506-6.

Pintrich, P. R. 2003. A motivational science perspective on the role of student motivation in learning and teaching contexts. *Journal of Educational Psychology*, 95, 667.

Pirani, S. I., & Arafat, H. A. 2016. Reduction of food waste generation in the hospitality industry. *Journal of Cleaner Production*, 132, 129–145.

Pisani, N., Kourula, A., Kolk, A., & Meijer, R. 2017. How global is international CSR research? Insights and recommendations from a systematic review. *Journal of World Business*, 52, 591–614.

Platonova, E., Asutay, M., Dixon, R., & Mohammad, S. 2018. The impact of corporate social responsibility disclosure on financial performance: evidence from the GCC Islamic banking sector. *Journal of Business Ethics*, 151, 451–471.

Rajasekar, J., & Simpson, M. 2014. Attitudes toward business ethics: a gender-based comparison of business students in Oman and India. *Journal of Leadership, Accountability and Ethics*, 11, 99.

Rettab, B., Brik, A. B., & Mellahi, K. 2009. A study of management perceptions of the impact of corporate social responsibility on organisational performance in emerging economies: the case of Dubai. *Journal of Business Ethics*, 89, 371–390.

Rice, G. 1999. Islamic ethics and the implications for business. *Journal of Business Ethics*, 18, 345–358.

Ryan, R. M., & Deci, E. L. 2000. Self-determination theory and the facilitation of intrinsic motivation, social development, and well-being. *American Psychologist*, 55, 68.

Saeed, S., & Zyngier, D. 2012. How motivation influences student engagement: a qualitative case study. *Journal of Education and Learning*, 1(2), 1–16.

Schmidt, M. A., & Cracau, D. 2017. A cross-country comparison of the corporate social responsibility orientation in Germany and Qatar: an empirical study among business students. *Business and Professional Ethics Journal*, 37(1), 67–104.

Schwartz, M. S. 2012. The state of business ethics in Israel: a light unto the nations? *Journal of Business Ethics*, 105, 429–446.

Setó-Pamies, D., & Papaoikonomou, E. 2016. A multi-level perspective for the integration of ethics, corporate social responsibility and sustainability (ECSRS) in management education. *Journal of Business Ethics*, 136, 523–538.

Smith, L. M. 2003. A fresh look at accounting ethics (or Dr. Smith goes to Washington). *Accounting Horizons*, 17, 47–49.

Spence, M., Gherib, J. B. B., & Biwolé, V. O. 2011. Sustainable entrepreneurship: is entrepreneurial will enough? A north–south comparison. *Journal of Business Ethics*, 99, 335–367.

Taleb, H. M. 2014. The potential for launching a postgraduate course on sustainable energy in Saudi Arabia. *Curriculum Journal*, 25, 432–458.

Trencher, G. P., Yarime, M., & Kharrazi, A. 2013. Co-creating sustainability: cross-sector university collaborations for driving sustainable urban transformations. *Journal of Cleaner Production*, 50, 40–55.

United Nations Global Compact. 2007. *The Principles for Responsible Management Education*, New York.

United Nations Principles for Responsible Management Education (UNPRME). 2019. Website. Available online at https://www.unprme.org/. Accessed May 2019.

Wafik, G. M., Fawzy, N. M., & Hassanein, F. A. 2011. The role of educational institutions in greening education and ensuring sustainability: the case study of faculty of tourism and hotels: Fayoum University. *International Journal of Hospitality & Tourism Systems*, 4.

Walters, W. H. 2009. Google Scholar search performance: comparative recall and precision. *portal: Libraries and the Academy*, 9, 5–24.

Wood, D. J. 1991. Corporate social performance revisited. *Academy of Management Review*, 16, 691–718.

Wright, T. S., & Wilton, H. 2012. Facilities management directors' conceptualizations of sustainability in higher education. *Journal of Cleaner Production*, 31, 118–125.

2 Corporate social responsibility (CSR) and the developing world

Highlights on the Egyptian case with implications for CSR education

Dina El-Bassiouny

Introduction

The enormous boost of technological advancements has unfortunately caused remarkable environmental damage. This is why "sustainable development" became an important issue for scholars to discuss. Advances in sustainability knowledge and concepts have increased as well as research on the social and environmental activities of corporations. Along the same line, recent corporate scandals have awakened public awareness and increased the need for corporate social responsibility (CSR) practices (Lee & Carroll, 2011). CSR disclosures have now become common practice by businesses, where companies provide CSR reports worldwide, and their number is growing every year (Ernst & Young & Boston College Center for Corporate Citizenship, 2016).

Yet, the diversity in CSR application in different societies and the reason for their existence are not adequately reflected in mainstream CSR research. Also, prior CSR literature is largely Western-driven. Accordingly, this chapter aims to make an important contribution by presenting an introspect into the unique setting of CSR in developing countries through a highlight of the Egyptian case. The chapter starts with presenting an overview of different conceptualizations of the CSR notion. This is followed by presenting different debates in the literature with respect to CSR. A focus on the literature dealing with developing countries then follows. The Egyptian case is then presented.

Different conceptualizations of CSR

Despite the widespread conceptualization and operationalization of CSR in the academic business literature (Matten & Moon, 2008), theoretical and empirical inconsistencies are still apparent in the meanings and practices of corporate responsibility (McWilliams et al., 2006). Several definitions have been proposed by academic researchers in an attempt to define and further develop the CSR construct. The most prominent of these is the definition developed by Carroll (1979) in his well-known research study "A three-dimensional conceptual model of corporate performance." In his social performance model, Carroll (1979) defines the social responsibility of businesses as "the economic, legal, ethical, and discretionary expectations that society has of organizations at a given point in time" (p. 500). According to this view, the economic and legal actions of corporations are considered part of their social responsibility. However, other authors argue that the economic performance of corporations is the main motive for their existence and should therefore not be considered as a responsibility of businesses towards their societies. According to Turker (2009), for instance, CSR is defined as "the corporate behaviors that aims to affect stakeholders positively and that go beyond its economic interest" (p. 413).

Some business scholars also questioned the inclusion of the legal component in CSR. McWilliams & Siegel (2001), for instance, view the social responsibilities of firms as "actions that appear to further some social good, beyond the interests of the firm and that which is required by law" (p. 117). Matten and Moon (2008) also argue that CSR at its core represents "clearly articulated and communicated policies and practices of corporations that reflect business responsibility for some of the wider societal good" and "is therefore differentiated from business fulfillment of core profit-making responsibility and from the social responsibilities of government" (p. 405). Accordingly, based on this view, the economic as well as the legal responsibilities of corporations are considered as regular corporate actions but not related to CSR.

Furthermore, Campbell (2007) argues that the corporate responsibilities offered by most CSR literature largely neglects corporate irresponsible behavior. As such, he presents a CSR definition that "sets a minimum behavioral standard with respect to the corporation's relationship to its stakeholders, below which corporate behavior becomes socially irresponsible" (p. 951). This minimal responsible behavior requires corporations to perform two main actions. First, not deliberately causing any harm to corporate stakeholders including employees, investors, customers and society. Second, rectifying discovered corporate harm whether voluntarily or as a result of normative or legal pressures. Accordingly, this definition developed by Campbell (2007) focuses on the minimum end of the continuum rather than corporate behavior located on the other end. In a broader understanding of the CSR continuum, Strike et al. (2006) distinguished between corporate responsible and irresponsible behavior. They define corporate irresponsible acts as "corporate actions that negatively affect an identifiable social stakeholder's legitimate claims" (p. 852). In their study, they argue that the social responsibility of corporations does not necessarily undo corporate irresponsible actions and have thus classified two separate constructs for CSR and criticized the aggregation of both CSR dimensions.

Based on this quick review of CSR definitions, it is obvious that the "CSR continuum" is wide and diverse. As Campbell (2007) puts it, "socially responsible corporate behavior may mean different things in different places to different people and at different times" (p. 950). Historical shifts (Campbell, 2007), cross-country differences (Chapple & Moon, 2005; Matten & Moon, 2008), and the different academic interpretations of CSR reduce the possible standardization of the CSR construct.

CSR and the developing world: a focus on Egyptian companies

The business and society debates

Different debates over the social responsibility of businesses have produced various schools of thought. On the one end, the followers of Milton Friedman view CSR as a "misguided" business activity that reduces profits while the main concern of businesses should be profit maximization (Blowfield & Frynas, 2005; Bowd et al., 2006). This view treats businesses as not having any social obligations as the business entities are not people, hence, these cannot be considered as citizens (Ofori & Hinson, 2007). On the other end, the pro-CSR school criticizes business behavior and argues that the survival of businesses is dependent upon the benefits they consume from their societies (Ofori & Hinson, 2007) and by that businesses have a responsibility towards societies (Bowd et al., 2006). Accordingly, businesses should be responsible for all the impacts of their activities not only on shareholders but also on the society at large (Blowfield & Frynas, 2005; Ofori & Hinson, 2007).

Within the pro-CSR school, various justifications exist for the CSR initiatives employed by companies. According to Porter and Kramer (2006), there are four arguments for the engagement of companies in CSR initiatives: moral obligation, sustainability, license to operate and reputation. The moral justification principle argues that companies should act as moral citizens and consider different stakeholders affected by their operations, including local communities and natural environment. The sustainability notion emphasizes business stewardship obligations towards future generations. The license to operate argues that corporate social activities are used as a way to legitimize corporate actions in the eyes of the different stakeholders, including governments and communities. The final justification for CSR, reputation, notes that companies initiate CSR activities to enhance corporate image and may thus even improve corporate stock value (Porter & Kramer, 2006).

However, the manifestation of the varying views revolving around the conceptualization and importance of CSR in developing countries remain unclear. Dobers and Halme (2009) argue that the prominent CSR views are mainly addressed in Western-driven literature, hence may not apply in a developing country context. The economic, social, and political concerns faced by many developing economies may create unique arguments for corporate engagement in CSR practices. In the general sense, operating in a relatively unhealthy economic environment that is characterized by high inflation rates, low productivity, and relative immaturity of consumers (and financial) markets causes uncertainties regarding short-term profit-yielding (Campbell, 2007). Accordingly, corporations operating in such environments tend to focus more on increasing their basic economic growth (Ofori & Hinson, 2007) rather than focusing on higher-level sustainability issues. Accordingly, and in the presence of weak institutional conditions, CSR is pursued mainly by cause of "personal discretion, hindsight, and initiative" (Jamali & Mirshak, 2007, p. 260).

Corporate social practices in developing countries are also often regarded as gap fillers where governments have failed. Given the limited role played by governments in the social domain of many developing countries (Jamali & Neville, 2011), a number of studies on developing countries highlight the social role of businesses in developing respective societies where governments fall short in performing their public duties (Amaeshi et al., 2006; Frynas, 2005; Ite, 2004). In the Middle East, the poor performance of governments in various countries represents one possible explanation for adoption of corporate social activities by businesses. Given the insufficient social services provided by governments in this region, it is argued that most philanthropic contributions made by corporations are aimed at supporting education, research, and health services to "reduce the social tension, gain legitimacy and create a safer environment to conduct business" (Ararat, 2006, p. 5).

The CSR literature in developing countries

On the theoretical level, the literature domain on CSR in developing countries is growing. In their comprehensive review of 452 CSR articles in developing countries from 1990 to 2015, Jamali and Karam (2016) found that 93% of the reviewed articles were published in the period between 2005 and 2015. This indicates the emergence of CSR as a domain of study in developing countries.

Their study also shows that the literature on CSR in developing countries exhibits high variation and is distinguishable due to five main aspects. First, given the specific and complex characteristics of the institutional environment in many developing countries, the literature on CSR in developing countries is most distinctive at the institutional level of analysis. The unstable or failing government systems, ineffective regulatory frameworks, high levels of corruption, and underdeveloped governance structures contribute to what is referred to as "institutional voids" in many developing countries (Jamali & Karam, 2016). Such institutional voids represent one type of regulatory void that indicate a lack of competent institutions that enforce regulations and norms. In the presence of weak regulatory environment, the embeddedness of societal concerns into corporate strategy and operations becomes very unlikely (Short, 2013), and hence, the proper application of CSR in such a context is highly questioned (Campbell, 2007). Also, the contradictory influence of the various institutions on promoting or obstructing corporate social behavior creates countering effects on the authenticity and structure of CSR practices. As such, analyses of CSR in developing countries reflect the different institutional conditions that shape unique conceptualizations and applications of CSR that can hardly be placed within predominant mainstream CSR literature (Jamali & Karam, 2016).

The second important differentiator of the CSR literature in developing countries represents the complex effects of other macro-level factors that extend beyond the business operating environment, including geo-political factors and globalization. A critical review of the literature clearly highlights existing tensions between corporate attempts to adopt best practices of CSR due to the spread of international organizations and standards and the complications and burdens placed on corporations by past geo-political factors. Third, the literature captures the complexities of governance systems in developing countries. The institutional voids of formal and informal governance systems amplify the space for agency by multiple actors who can strategically pursue their interests by taking advantage of encountered voids. In the presence of weak governmental influence, other actors, including political elites and local and multinational firms, may intervene and assume active roles of governance (Jamali & Karam, 2016). At the organizational level, it is argued that publicly listed companies in developing countries often adopt the external image of corporate governance practices rather than the core substance (Peng, 2004). Local companies can take advantage of informal networks to reduce formal governance control while continuing to pursue philanthropic contributions (Jamali & Karam, 2016). Having ineffective corporate governance mechanisms leads to more agency problems (Harjoto & Jo, 2011) that can have salient implications on CSR application. Accordingly, within this specific institutional environment, the dynamics of governance in developing countries and its CSR implications can be better comprehended if the various non-traditional actors and the roles they play are clearly identified (Jamali & Karam, 2016).

The fourth aspect deals with the varied forms of CSR expressions across developing countries. It can be said that different forms of CSR are shaped depending on the varying effects of multiple socio-political and historical factors and that explicit global forms of CSR are mainly imitated to serve specific purposes. The varied CSR expressions in developing countries can be better understood if the countervailing effects of the complex institutional environment and the existence of multiple governance actors are taken into account (Jamali & Karam, 2016). The final aspect is related to CSR consequences in developing countries as highlighted by previous studies (Karam & Jamali, 2013; Kolk & Lenfant, 2013). A wide scope of consequences is depicted and varies between financial and social outcomes as well as positive and negative considerations regarding individuals and the society at large (Jamali & Karam, 2016).

Those aspects highlight the unique settings of developing countries and how they contribute to distinguishable research in the CSR field within developing countries as well as worldwide. Understanding where the literature on CSR in developing countries currently stands initiates a starting point from which future research on CSR in developing countries can begin.

Background on the Egyptian business climate

Egypt is a large populous country in the Middle East with a total area of one million square kilometers and a population of around 91.5 million inhabitants (The World Bank, 2016a). Since 2011, a major political and social transition has been witnessed as well as considerable economic disruptions that have negatively influenced various sectors in Egypt, including tourism, retail, and banking (The World Bank, 2013).

According to Enterprise surveys conducted by the World Bank in 2013 on 2,897 firms operating in Egypt, business managers ranked political instability as the biggest obstacle faced by the private sector in Egypt. Despite obvious challenges created by the 2011 political event, this transitional change can still be positively utilized by policymakers to sustainably address major challenges and obstacles embedded in the Egyptian business environment. The presence of anti-competitive practices, corruption, and regulatory policy uncertainty represent major challenges that hinder private sector activity. Enterprise surveys show that the private sector in Egypt identified practices of the informal sector among the top 10 obstacles to doing business. Almost 48% of firms are competing against informal firms in Egypt. This high level of informality in the market is an outcome of excessive and complicated regulations and administrative procedures for starting a business in Egypt (The World Bank, 2013). The rank of Egypt in the World Bank's Doing Business 2016 Report was 131st (The World Bank, 2016b) implying a complex business environment. Applying easier and more transparent rules would assist in increasing the number of formal firms, enlarging the tax base, and reducing corruptive practices. Along the same line, reducing the gap between the written and the applied law would pave the way to more equitable economic growth and more job opportunities across Egypt (The World Bank, 2013).

In developing countries, the private sector provides an estimated 90% of jobs, making it a major employer. Given the high and rising unemployment rates in Egypt (13%), specifically amongst the youth, the role of the private sector to provide quality formal jobs is becoming more and more crucial. Yet, the vigorous growth of the private sector, and the economy at large, is dependent upon the application of sound business regulations. In Egypt, the central government conducts most of the government's business and a concentration of power exists among superior government actors. To streamline policy implementation and increase accountability, more power delegation to administrative units is required. A redefinition of institutional roles and processes and coordination among the different sectors and institutions can increase government responsiveness to local private sector needs (The World Bank, 2013).

The philosophy of CSR existence in Egypt

To explore the main drivers for CSR involvement in Egypt, in-depth interviews with managers in Egyptian companies were conducted. The concept of CSR was introduced to managers as the responsibility of the company towards its stakeholders, specifically the community, natural environment, employees, customers, and the government. Yet, emphasis was given to the community aspect of CSR. The sample consisted of 109 companies listed in the Egyptian stock exchange located in Cairo, the capital of Egypt, and Alexandria, the second largest city in Egypt. Corporate managers from different service and industrial sectors were interviewed. The final sample consisted of 74 managers interviewed over a three-month period. The interviewees were asked about the underlying motives of Egyptian companies to perform CSR practices and the main reason behind the weak CSR initiatives in Egypt. Despite scholarly arguments on business engagement in CSR practices as a result of government failures, only 10.8% of respondents see that government failures are the reason for CSR practices in Egypt, as shown in Table 2.1. On the other hand, 44.6% of the respondents view publicity as the main driver of CSR in Egypt, followed by stakeholder pressure (28.4%) and globalization (13.5%). Reports on CSR in the Middle East indicate that society places more value on corporate image and reputation than observed corporate behavior (Ararat, 2006). In light of this fact, Egyptian corporations may use CSR as a tool to boost corporate image.

The majority of respondents (82.4%) also see that the main reason behind the lack of CSR initiatives in Egypt is the poor economic conditions of the country, while only 10.8% consider that lack of awareness and media pressure hinders social responsibility practices amongst Egyptian firms.

Also, companies operating in Egypt tend to reflect the stockholder view of CSR that represents a "limited and narrow view of corporate obligations only to stockholders" (Shafer et al., 2007, p. 270). This is apparent from our

Table 2.1 CSR drivers in Egypt

Drivers	%
Public relations	44.6
Stakeholder pressure	28.4
Globalization	13.5
Government failures	10.8
Corporate affiliation to society	8
Political reasons	4
International competitiveness	1.3
Government regulations	1.3

Table 2.2 Demographic characteristics of participants

Description	*Percentage*
Experience	
Less than 5 years	1.8
5 to 9 years	10
10 to 20 years	69.1
More than 20 years	19.1
Age	
25 to 34	5.5
35 to 44	34.5
45 to 55	53.6
Above 55	6.4
Education	
Bachelor's	15.5
Master's	70
Doctoral	14.5

survey results of 110 managers in a sample of 110 local and multinational companies operating in Egypt, who were surveyed over a three-month period about managerial attitudes toward social responsibility[1] further to the in-depth interviews conducted. Table 2.2 provides a detailed description of the characteristics of the participants. The 110 sample companies are selected based on a purposive criterion sampling method, where only companies with disclosed CSR practices were chosen. The sample is restricted to listed local companies in the Egyptian stock exchange and multinational companies operating in two or more countries and with headquarters based outside of Egypt. Based on that, the final sample consists of 54 local and 56 multinational companies. The sample includes companies from diverse industries such as manufacturing, pharmaceutical, consulting and banking industries.

As shown in Figure 2.1, around 74% of companies do not agree with the notion that a firm's ethical and social responsibility is essential to its long-term profitability. Also, 76.4% of the companies surveyed disagree that businesses have a social responsibility beyond making a profit, and around 72% of the companies surveyed confirm the statement that "the most important concern for a firm is making a profit, even if it means bending or breaking the rules." On the other hand, a small percentage of managers agree with the first two statements (1% and 26.4% respectively) that resemble the stakeholder view of CSR.

This all provides insights as to how CSR is viewed and practiced in Egypt. It is well obvious that the practice of CSR amongst firms operating in Egypt is mainly viewed as a tool used by companies to improve corporate branding and reputation. Also, the economic conditions of the country represent a main obstacle to the proper initiation of CSR practices in Egypt. Underneath

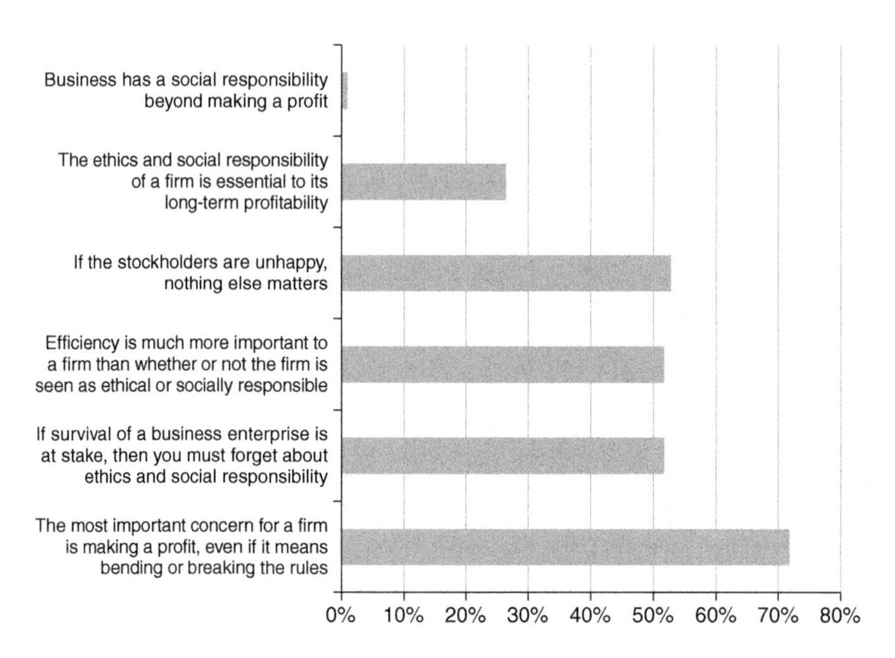

Figure 2.1 Managerial attitudes towards social responsibility in Egypt.

such views lie two main arguments. First, less profitable firms are less likely to engage in CSR practices compared with more profitable firms due to the lack of resources required for such activities (Campbell, 2007). Many research studies on the association between firm profitability and CSR have revealed positive associations between corporate financial and social performance (Margolis & Walsh, 2003). Second, to maintain short-term profit maximization, firms operating in a weak economic and regulatory environment with relatively low stakeholder pressure and awareness are less likely to act in socially responsible ways (Campbell, 2007). In Egypt, regulations controlling corporate social and environmental behavior exist. Yet, the enforcement of these regulations is rare and selective, resulting in a gap between written laws and practice. The presence of weak and ineffective enforcing bodies produces inconsistent enforcement and inspection and results in corporate non-compliance with laws. In addition, the low level of disclosure practices that address social concerns pose questions regarding the ethical stances of Egyptian companies (Hanafi, 2006).

Barriers to effective CSR implementation in Egypt

Common CSR barriers in developing countries include ineffective regulatory and governance systems, relatively high levels of corruption, lack of top management commitment, and insufficient levels of CSR expertise (Arevalo & Aravind, 2011; Nasrullah & Rahim, 2014). These factors combined reduce corporate tendency to act in the public's interest and curtail their commitment to effective CSR implementation (Nasrullah & Rahim, 2014). Although CSR is gaining some momentum among private companies in Egypt, many restraints hinder its successful implementation.

To explore the challenges of integrating CSR practices in the Egyptian context, the 110 managers from local and multinational firms, referred to in the previous section, were surveyed to explore the barriers for implementing social and environmental improvement in their companies.[2] Figure 2.2 shows some of the factors that may limit the implementation of CSR in corporate business units and the percentage of managers that perceive those factors as relevant barriers in Egypt. The majority of managers (67.3%) consider the lack of top management commitment as the main barrier for adopting CSR in their corporation. This indicates that CSR issues are still a second-tier concern for corporate elites in Egypt. It also shows that top management plays an important role in launching and promoting CSR implementation in corporations operating in Egypt. Also, 41.8% of managers consider the lack of customer demand as an important barrier to CSR implementation. Devoting resources to CSR necessitates extant pressures from multiple stakeholder groups, including customers (McWilliams & Siegel, 2001). If stringent customer demands are sought, raising customer awareness on social and environmental issues seems salient for fostering CSR practices in Egypt.

Around 35% of the managers consider the lack of employee interest in CSR and competition on costs as important barriers for CSR implementation in Egypt. Employees are important stakeholders who can play an important role in pressuring companies to adopt work-related CSR policies (McWilliams & Siegel, 2001). Recognizing employee demand for CSR,

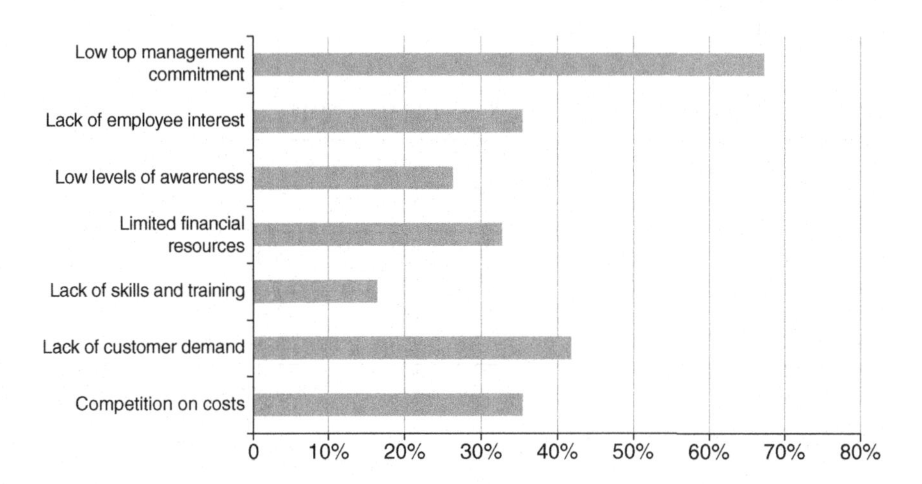

Figure 2.2 Barriers for CSR in Egypt.

managers are more likely to support work-related CSR issues such as financial security and safety. On the other hand, low levels of awareness of CSR issues (26.3%) and lack of skills and training (16.4%) are not considered as relevant barriers by the majority of managers surveyed. Overall, multi-stakeholder engagement, specifically employees and customers, top management commitment, and going above and beyond market-based success criteria are considered important contributors to the promotion of CSR in Egypt.

Strategic CSR stances: where does Egypt stand?

Corporate social responsiveness represents one important component in the social performance of corporations. This dimension addresses "the philosophy, mode, or strategy behind business (managerial) response to social responsibility and social issues" (Carroll, 1979, p. 501). Businesses respond to social issues and pressures in different ways, ranging from having no response to being a proactive leading company in the industry. Four main categories of social responsiveness have been identified by previous CSR literature: reactive (obstructionist), defensive, accommodative, and proactive (Carroll, 1979; Henriques & Sadorsky, 1999; Lee, 2011). The degree to which corporations respond to social issues is dependent upon the strength of the institutional and stakeholder influences surrounding the business environment. Lee (2011) provides two interdependent social mechanisms that represent the external influences that shape the CSR strategies used by corporations in response to social issues. First, corporate social behavior, in general, is affected by the strength of the regulative, normative, and cognitive institutional influences. And second, stakeholder influences can pressure corporations to serve stakeholder interests and other social demands. Depending on the strength and alignment of the signals provided by those two mechanisms on CSR-related issues, corporations create CSR strategies equivalent to the relative reinforcement of such issues by external influences.

Table 2.3 Level of CSR commitment of companies in Egypt

CSR commitment	*Committed*		*Do not know*		*Uncommitted*	
	No. of firms	*%*	*No. of firms*	*%*	*No. of firms*	*%*
CSR 0lan	56	50.9	23	20.9	31	28.2
Written plan	20	18.2	37	33.6	53	48.2
Plan communicated to	21	19.1	45	40.9	44	40
shareholders	41	37.3	27	24.5	42	38.2
Plan communicated to	41	37.3	13	11.8	56	50.9
employees	28	25.5	41	37.3	41	37.3
CSR unit	85	77.3	13	11.8	12	10.9
External CSR reporting						
Employee CSR training						

Table 2.3 shows managers' perceptions on the CSR commitment of a sample of 110 local and multinational companies operating in Egypt. This was examined based on seven main aspects: (1) having a CSR plan, (2) having a formal document describing the plan, (3) presenting the plan to shareholders and stakeholders, (4) presenting the plan to employees, (5) having a CSR unit dealing with CSR issues, (6) providing external CSR reporting, and (7) providing employee CSR training (Henriques & Sadorsky, 1999). While around 50% of firms examined have a CSR plan, only a small percentage of firms have it in a written form (18.2%) and communicate it to their shareholders (21%). It is also obvious that the extent of external CSR reporting amongst examined firms is still immature, where only 25.5% confirmed the presence of external CSR information disclosure. On the other hand, employee CSR training seems to be well established, as almost 77.3% of examined firms provide CSR training to their employees. The weak communication of the CSR plan and CSR information to shareholders and various stakeholders suggests the lack of effective communication channels with corporate stakeholders and ineffective strategic stakeholder management amongst the sampled Egyptian companies.

In Egypt, like many other developing countries, the weak regulation enforcement (Dobers & Halme, 2009; Graham & Woods, 2006; Qu, 2007) and low awareness or demand for CSR (Amran & Devi, 2008; Naser et al., 2006) create challenging CSR environments that may hinder the application of CSR practices in developing countries and discourage the embeddedness of CSR in corporate strategies. This argument is consistent with Lee's (2011) proposed model that suggests that companies operating in an environment where institutional and stakeholder pressures are weak tend to follow the obstructionist (reactive) approach to CSR. Corporations employing such a CSR strategy reject social responsibilities that lie outside their economic interests (Lee, 2011) and do not perceive CSR issues as a priority (Henriques & Sadorsky, 1999). In such an environment, companies lack the incentive to act in a socially responsible manner and may even bear a competitive disadvantage from acting responsibly. Under such conditions, the obstructionist (reactive) strategy is the expected CSR strategy for companies.

Conclusion and implications for CSR awareness and education

The studies conducted by authors and reported here highlight the status of CSR in Egypt, a Middle Eastern developing country. The drivers, barriers, and degree of CSR commitment, as perceived by corporate managers operating in Egypt, reveal that the conceptualization and operationalization of CSR is still relatively underdeveloped in the Egyptian market. The study shows that CSR in Egypt may be viewed as a shallow short-term approach that can be used to boost corporate reputation. This implies that CSR is still not embedded as a long-term corporate strategy to enhance the social, environmental as well as the economic impact of corporate activity. One of the main external challenges faced by corporations is the existence of an economic and institutional environment that hinders the development of CSR in Egypt. This includes the relatively unfavorable economic conditions that pressure companies to focus on maintaining short-term profits as well as the existence of weak and ineffective enforcement of the law.

From an internal managerial perspective, the studies reveal that the two major barriers of CSR implementation in Egypt are the low levels of top-management commitment and lack of customer demand for CSR. This implies the essential need for enhancing CSR programs for awareness and education on both the business and community levels. To achieve higher levels of sustainable societies, public awareness, training and education are essential tools that need to be utilized to reach that end (Hopkins and Mckeown, 2002).

On the business level, previous studies show that awareness of managers about sustainability management tools is considered to be one of the most important drivers of adopting and implementing sustainability management tools in business organizations (Johnson, 2015; Schaltegger et al., 2012). A diverse number of tools are suggested by previous literature including environmental and social audits, sustainability-oriented accounting tools, and employee development tools. The purpose of sustainability management tools is to develop and improve internal social and environmental corporate performance as well as external communication with stakeholders on sustainability issues (Johnson, 2015). Given its critical importance, it is argued that the first step for embedding sustainability in organizations is raising awareness, or else the adoption of sustainability management tools would be very unlikely (Johnson, 2015). Schaltegger et al. (2012) mention that a key driver for the application of sustainability management tools in organizations, industries, and countries is the acquisition of knowledge about such tools. Accordingly, increasing awareness of sustainability approaches and tools, whether existing or new, can contribute to increasing the operationalization of sustainability tools and thus enhancing corporate sustainability (Schaltegger et al., 2012).

On the community level, awareness is also considered to be one of the main tools that can be used to promote sustainable development in societies. It is argued that sustainable development cannot be achieved unless there is awareness of the interrelationship between the economic, environmental, and social aspects of the society amongst citizens and specialists (Erdogan & Tuncer, 2009) and the public community is actively engaged in building a sustainable future (Barr, 2003). To effectively develop awareness and achieve sustainability, a culture for sustainable development has to be cultivated (Erdogan & Tuncer, 2009; Hawkes, 2001). This can be done through embedding a sustainable culture in different social entities, including education and training, communications, arts, leisure, and sports (Hawkes, 2001).

Universities, being a common example of formal education, are considered one of the major drivers of change that can cultivate a culture of sustainability in societies (AlShuwaikhat et al., 2016). Higher education for sustainable development can be seen as an essential tool that can help individuals of the society to build, enhance, and transfer knowledge on sustainable development as well as realize the broad effects of human behavior and decisions on sustainability. As such, the role of academia in promoting sustainable development should encompass discussions on ways for creating a sustainable lifestyle, whether in the individual's personal life or at work, and include not only formal learning processes, such as study programs and courses, but also informal learning techniques, such as student volunteering (Barth et al., 2007). University policies can also be conducted with the aim of enhancing the environmental literacy of faculty and staff members of universities, embedding sustainability concepts in the curricula of universities and implementing environmental management systems that perform input-output analyses for sustainability. Such sustainability policies are deemed helpful in facilitating change toward embedding sustainability into academic institutions (Wright, 2002).

Increasing awareness and building a culture of sustainability on both the business and community level seems vital for the operationalization and institutionalization of sustainability tools and concepts within organizations. This is of specific importance in Egypt and other countries with similar contexts where an enabling environment for CSR application is lacking. If sustainability awareness and culture are cultivated, organizations would move to a more effective level of CSR operationalization where sustainability will be more deeply embedded in organizational strategy and the positive impact of such practices can significantly contribute to more sustainable societies.

Notes

1 Survey items of managerial attitudes toward social responsibility are adopted from Shafer et al. (2007).

2 Survey items for barriers for implementing CSR are adopted from Pedersen and Neergaard (2009).

References

Alshuwaikhat, H., Adenle, Y., & Saghir, B. (2016). Sustainability assessment of higher education institutions in Saudi Arabia. *Sustainability*, *8*, 750.

Amaeshi, K., Adi, B., Ogbechie, C., & Amao, O. (2006). Corporate social responsibility in Nigeria: Western mimicry or indigenous influences? *Journal of Corporate Citizenship*, *24* (Winter), 83–99.

Amran, A., & Devi, S. (2008). The impact of government and foreign affiliate influence on corporate social reporting: The case of Malaysia. *Managerial Auditing Journal*, *23*(4), 386–404.

Ararat, M. (2006). *Corporate Social Responsibility across Middle East and North Africa.* Turkey: Sabanci University.

Arevalo, J., & Aravind, D. (2011). Corporate social responsibility practices in India: Approach, drivers, and barriers. *Corporate Governance: The International Journal of Business in Society, 11* (4), 399–414.

Barr, S. (2003). Strategies for sustainability: Citizens and responsible environmental behaviour. *Area, 35* (3), 227–240.

Barth, M., Godemann, J., Rieckmann, M., & Stoltenberg, U. (2007). Developing key competencies for sustainable development in higher education. *International Journal of Sustainability in Higher Education, 8* (4), 416–430.

Blowfield, M., & Frynas, J. (2005). Setting new agendas: Critical perspectives on corporate social responsibility in the developing world. *International Affairs, 81* (3), 499–513.

Bowd, R., Bowd, L., & Harris, P. (2006). Communicating corporate social responsibility: An exploratory case study of a major UK retail centre. *Journal of Public Affairs, 6*, 147–155.

Campbell, J. (2007). Why would corporations behave in socially responsible ways? An institutional theory of corporate social responsibility. *The Academy of Management Review, 32* (3), 946–967.

Carroll, A. (1979). A three-dimensional conceptual model of corporate performance. *The Academy of Management Review, 4* (4), 497–505.

Chapple, W., & Moon, J. (2005). Corporate social responsibility (CSR) in Asia: A seven-country study of CSR web site reporting. *Business and Society, 44* (4), 415–441.

Dobers, P., & Halme, M. (2009). Editorial: Corporate social responsibility and developing countries. *Corporate Social Responsibility and Environmental Management, 16*, 237–249.

Erdogan, M., & Tuncer, G. (2009). Evaluation of a course: "Education and Awareness for Sustainability." *International Journal of Environmental and Science Education, 4* (2), 133–146.

Ernst & Young (EY), & Boston College Center for Corporate Citizenship (2016). *Value of Sustainability Reporting*. Retrieved March 2016 from www.ey.com/US/en/Services/Specialty-Services/Climate-Change-and-Sustainability-Services/Value-of-sustainability-reporting.

Frynas, J. (2005). The false developmental promise of corporate social responsibility: Evidence from multinational oil companies. *International Affairs, 81* (3), 581–598.

Graham, D., & Woods, N. (2006). Making corporate self-regulation effective in developing countries. *World Development, 34* (5), 868–883.

Hanafi, R. (2006). *An Exploration of Corporate Social and Environmental Disclosure in Egypt and the UK: A comparative study*. PhD thesis. University of Glasgow.

Harjoto, M., & Jo, H. (2011). Corporate governance and CSR nexus. *Journal of Business Ethics, 100*, 45–67.

Hawkes, J. (2001). *The Fourth Pillar of Sustainability: Culture's essential role in public planning*. Australia: Common Ground Publishing.

Henriques, I., & Sadorsky, P. (1999). The relationship between environmental commitment and managerial perceptions of stakeholder importance. *The Academy of Management Journal, 42* (1), 87–99.

Hopkins, C., & Mckeown, R. (2002). Education of sustainable development: An international perspective. In D. Tilbury, R. Stevenson, J. Fien, & D. Schreuder (eds.), *Education and Sustainability: Responding to the Global Challenge*. Switzerland and UK: IUCN.

Ite, U. (2004). Multinationals and corporate social responsibility in developing countries: A case study of Nigeria. *Corporate Social Responsibility and Environmental Management, 11*, 1–11.

Jamali, D., & Karam, C. (2016). CSR in developing countries as an emerging field of study. *International Journal of Management Reviews, 0* (0), 1–30.

Jamali, D., & Mirshak, R. (2007). Corporate social responsibility (CSR): Theory and practice in a developing country context. *Journal of Business Ethics, 72* (3), 243–262.

Jamali, D., & Neville, B. (2011). Convergence versus divergence of CSR in developing countries: An embedded multi-layered institutional lens. *Journal of Business Ethics, 102*, 599–621.

Johnson, M. (2015). Sustainability management and small and medium-sized enterprises: Managers' awareness and implementation of innovative tools. *Corporate Social Responsibility and Environmental Management, 22*, 271–285.

Karam, C., & Jamali, D. (2013). Gendering CSR in the Arab Middle East: An institutional perspective. *Business Ethics Quarterly, 23* (1), 31–68.

Kolk, A., & Lenfant, F. (2013). Multinationals, CSR and partnerships in Central African conflict countries. *Corporate Social Responsibility and Environmental Management, 20* (1), 43–54.

Lee, M. (2011). Configuration of external influences: The combined effects of institutions and stakeholders on corporate social responsibility strategies. *Journal of Business Ethics, 102* (2), 281–298.

Lee, S., & Carroll, C. (2011). The emergence, variation, and evolution of corporate social responsibility in the public sphere, 1980–2004: The exposure of firms to public debate. *Journal of Business Ethics, 104*, 115–131.

Margolis, J., & Walsh, J. (2003). Misery loves companies: Rethinking social initiatives by business. *Administrative Science Quarterly, 48* (2), 268–305.

Matten, D., & Moon, J. (2008). "Implicit" and "explicit" CSR: A conceptual framework for a comparative understanding of corporate social responsibility. *The Academy of Management Review, 33* (2), 404–424.

McWilliams, A., & Siegel, D. (2001). Corporate social responsibility: A theory of the firm perspective. *The Academy of Management Review, 26* (1), 117–127.

McWilliams, A., Siegel, D., & Wright, P. (2006). Guest editors' introduction: Corporate social responsibility: Strategic implications. *Journal of Management Studies, 43* (1), 1–18.

Naser, K., Al-Hussaini, A., Al-Kwari, D., & Nuseibeh, R. (2006). Determinants of corporate social disclosure in developing countries: The case of Qatar. *Advances in International Accounting, 19*, 1–23.

Nasrullah, N., & Rahim, M. (2014). CSR in private enterprises in developing countries: Evidence from the ready-made garments industry in Bangladesh. In S. Idowu, & R. Schmidpeter (eds.), *CSR, Sustainability, Ethics and Governance*. Switzerland: Springer.

Ofori, D., & Hinson, R. (2007). Corporate social responsibility (CSR) perspectives of leading firms in Ghana. *The International Journal of Business in Society, 7* (2), 178–193.

Pedersen, E., & Neergaard, P. (2009). What matters to managers? The whats, whys, and hows of corporate social responsibility in a multinational corporation. *Management Decision, 47* (8), 1261–1280.

Peng, M. (2004). Outside directors and firm performance during institutional transitions. *Strategic Management Journal, 25* (5), 453–471.

Porter, M., & Kramer, M. (2006). Strategy and society: The link between competitive advantage and corporate social responsibility. *Harvard Business Review,* (December), 1–13.

Qu, R. (2007). Corporate social responsibility in China: Impact of regulations, market orientation and ownership structure. *Chinese Management Studies, 1* (3), 198–207.

Schaltegger, S., Windolph, S., & Herzig, C. (2012). Applying the known: A longitudinal analysis of the knowledge and application of sustainability management tools in large German companies. *Society and Economy, 34* (4), 549–579.

Shafer, W., Fukukawa, K., & Lee, G. (2007). Values and the perceived importance of ethics and social responsibility: The U.S. versus China. *Journal of Business Ethics, 70,* 265–284.

Short, J. (2013). Self-regulation in the regulatory void: "Blue moon" or "bad moon"? *The ANNALS of the American Academy of Political and Social Science, 649* (1), 22–34.

Strike, V., Gao, J., & Bansal, P. (2006). Being good while being bad: Social responsibility and the international diversification of US firms. *Journal of International Business Studies, 37* (6), 850–862.

The World Bank (2013). *Doing Business in Egypt 2014: Understanding regulations for small and medium-size enterprises.* Available from www.doingbusiness.org.

The World Bank (2016a). *Egypt Home: Country at a glance.* Available from www.worldbank.org/en/country/egypt.

The World Bank (2016b). *Doing Business 2016: Measuring regulatory quality and efficiency, economy profile, Egypt, Arab Repub.* Available from www.doingbusiness.org/reports.

Turker, D. (2009). Measuring corporate social responsibility: A scale development study. *Journal of Business Ethics, 85* (4), 411–427.

Wright, T. (2002). Definitions and frameworks for environmental sustainability in higher education. *Higher Education Policy, 15,* 105–120.

3 Approaching the Giving Voice to Values (GVV) pedagogy in business ethics education

The case of the business ethics course at the German University in Cairo (GUC), Egypt

Noha El-Bassiouny, Sara Hamed, Nesma Ammar, Hadeer Hammad and Hagar Adib

Introduction

Ethics education is timely! With the compounded effects of ethical scandals such as Enron in 2001, Etihad Etisalat or Mobily in 2013 (Kerr, 2014), Volkswagen in 2019 (Welch, 2019), and global ethical fiascos such as the financial crisis of 2008, among others, ethics education becomes an imperative. Due to the escalating effects of lack of business ethics in the public sector, a global body of scholarship has also related business ethics to leadership (Downe et al., 2016). The question, then, becomes how to teach ethics, if ethics can be taught. Research tells us that ethics instruction can be effective (Wang & Calvano, 2015), yet because business ethics has roots in philosophy (Christensen et al., 2007; Mintz, 1996), the way it is taught can in fact become confusing. With most business ethics courses focusing on teaching through morally reasoning out ethical dilemmas (Trevino, 1992; Trevino & Nelson, 2011), ethical relativism can indeed cause confusion to students seeking better business ethics education.

Recently, however, an approach has emerged that advocates for certainty in business ethics education. This approach is the Giving Voice to Values (GVV) approach. The GVV philosophy is based on moral certainty, i.e., that the protagonist of the ethical situation knows the right thing to do and is actively seeking ways to reach how to do it (Gentile, 2012). The approach has been piloted globally and is creating a paradigm shift in how ethics instruction can move forward.[1] In 2015, it was piloted in Egypt for the first time through an international workshop that hosted students, instructors/educators, corporations, and community organizations in a co-learning approach. The result was the production of an Egyptian suite of GVV cases for use by instructors in Egypt and globally. These cases were integrated in a business ethics course targeting postgraduate MBA students.

The aim of this chapter is to showcase the transition of the Business Ethics course at the Faculty of Management Technology, the German University in Cairo, from one focused on ethical dilemmas to one focusing on teaching through GVV case studies. The chapter will start by reviewing international literature on business ethics, CSR, and sustainability education, then will move on to highlighting some Middle Eastern context issues. The course at the GUC will then be presented. This will be followed by highlighting the approach towards the GVV methodology and finally concluding remarks will be put forward for educators interested in piloting the course at their institutions.

Business ethics education: international perspectives

The transcendence of corporate scandals that took place in the last several decades had grasped considerable public attention related to shortcomings in ethical decision-making associated with these events while at the same time led to questioning the nature and effectiveness of business ethics educational approaches (Simola, 2014a). Developing more ethical managers is one of the prevailing challenges for business educators. In response to this need, most universities require students to complete a course in business ethics (Lam, 2004; Lawter et al., 2014). Western business schools have put great emphasis on ethics, yet the effectiveness of their programs are in question due to the failure of many business-school professional graduates, especially MBA graduates. Questions such as, 'do business ethics courses have any influence on the workplace environment?' and 'to what extent does it fit the capitalist economy?' need to be answered carefully by business educators (Lam, 2004).

Business ethics education has recently been acknowledged as an imperative element in business curricula at undergraduate and graduate programs alike (Wang & Calvano, 2015). The Association to Advance Collegiate Schools of Business (AACSB) has identified four universal domains that represent the underpinning of a holistic business ethics education curriculum. These domains include ethical decision-making, ethical leadership, corporate social responsibility and corporate governance (Simola, 2010). In recent years, business ethics education has advanced from a mere focus on ethical decision-making to cover two novel important areas: corporate social responsibility and sustainability education. Business ethics education focuses on fostering concepts of giving back to the society as well as preserving natural resources for future generations (Christensen et al., 2007). The AACSB, being the most highly recognized rating organization for the accreditation of business schools, advocates the inclusion of ethics education in business school curricula and supports the adoption of ethical decision-making, sustainability, and social responsibly issues. However, there is no rigid curricula that must be followed in teaching business ethics. A flexible approach in the administration of business ethics education is adopted based on the mission and learning objectives of each program (Floyd et al., 2013). Business school programs, hence, either adopt stand-alone courses or integrated courses where teaching ethics is combined as a component of other subjects. Recent research supports that around 70% of top business schools adopt ethics education, either as stand-alone or integrated courses (Neill, 2017; Wang & Calvano, 2015), while less than one-third of accredited business schools adopt stand-alone courses in teaching ethics for undergraduate or graduate programs (Floyd et al., 2013; Neill, 2017).

Rest's (1986, 1994) four-component framework of ethical decision-making and action represents one of the main foundations for behavioral frameworks used in business ethics (cf. Simola, 2014b). It involves early recognition of ethical concerns, evaluation or judgment to determine a good or "right" course of action, the formation of intention to give precedence to moral values, and the implementation of ethical action (Simola, 2014b). Another prevailing model in the literature of business ethics is Jones's (1991) Issue-Contingent Model, which combined prior ethical decision-making models and is argued to be mainly founded on Rest's (1986) model (cf. Craft, 2013). Jones (1991) introduced the concept of moral intensity, which consists of six components: (1) magnitude of consequences: the total harm/benefits of the moral act to the involved stakeholders; (2) social consensus: the level of social acceptance that a specific/proposed act is good or bad; (3) probability of effect: the possibility and extent the act will actually take place and will harm/benefit involved stakeholders; (4) temporal immediacy: time span between the present and the act; (5) proximity: the feeling of closeness to those involved; and (6) concentration of effect: the power of consequences for involved stakeholders. Generally, ethical decision-making is said to be influenced by three main categories of factors, which are individual factors (such as personality, personal values, and gender), organizational factors (education, experience, and decision style), and moral intensity (Craft, 2013).

Concerns are frequently being raised about the most effective means for teaching ethical decision-making and moral behavior (Floyd et al., 2013). Business ethics education has intensively followed cognitive approaches and methods in teaching ethics to students. Cognitive approaches focus on the communication of knowledge and moral reasoning through emphasizing ethical theories and frameworks along with their fundamental assumptions. Cognitive approaches accentuate on increasing the moral awareness and understanding of students about the ethical guidelines and principles to assist in increasing their ability to follow ethical decision-making in challenging situations. On the basis of ethical theories, positive and negative business ethical cases are presented and analyzed with students. Cognitive thinking stimulates students to consider the impact of decision-making on all interested stakeholders. In spite of the passive learning nature of cognitive approaches, these teaching techniques are expected to play a vital role in shaping the ethical systems of students (Chavan & Carter, 2018). However, the use of theoretical approaches in teaching business ethics is criticized for the lack of addressing the affective perspective of the learning process. Cognitive approaches might result in a mere knowledge acquisition without any emotional connection to decision-making. Contrary to cognitive approaches, affective teaching approaches are proclaimed to contribute to the development of moral sensitivity and ethical behavior (Brook & Christy, 2013). Experiential learning and action learning models, as affective teaching approaches, help in closing the gap between theory and practice and emphasize practical action in the workplace. Such models allow for the transfer and application of relevant knowledge and the change in ethical development of students through the use of discussions, real-life case studies, movies, role-plays, and simulations. The efficacy of affective approaches stem from its ability to increase the students' emotional involvement, question held beliefs, and create a milieu for resolving ethical conflict to help in the development of moral sensitivity and moral character (Chavan & Carter, 2018).

Business ethics class serves as a formal learning space for students. It exposes students to a wide variety of ethical dilemmas in organizations and societies, where the main objective is to enhance the students' moral awareness and maturity. Through engaging students with their personal values, ethics education enables students to correctly respond to ethical dilemmas that they are more likely to confront once they enter the business world. The courses are mainly designed to expose students to different types of ethical dilemmas by discussing theories as frameworks to assess potential outcomes. The case-based examples are then used to enable students to practice their newly acquired awareness. Yet, this approach is sometimes criticized for its lack of transferability to work experiences after university (Lawter et al., 2014). Recent research suggests four main segments for effective ethical decision-making course content. The first segment is related to ethical awareness or recognition, in which students are expected to be able to recognize/highlight a wide range of factors – from individual to organization level – that prohibit their ability to identify emerging ethical dilemmas/problems. The second segment is concerned with ethical evaluation or judgment skills. Through highlighting and discussing conventional philosophical theories, such as consequentialism, deontology, and virtue ethics as well as contemporary approaches related to care ethics, students are expected to advance into creating more creative ethical alternatives instead of just being able to identify ethical scenarios. The third segment focuses on the formation of ethical intention through prioritizing specific ethical values. Students are expected to shift from a reactive movement towards ethical behavior to a more proactive movement based on moral motivation. The fourth – and the final – segment is concerned with the implementation of ethical action through emphasizing the factors that either hinder or facilitate one's ability to apply ethical decisions. Topics discussed in the segments include but are not limited to diffusion of responsibility and consequences of certain compensation practices on ethical behavior (Simola, 2014b).

Although course content is considered a crucial element in ethics education, learning style is considered as important for the success of ethics education. Usually both researchers and educators emphasize the importance of how teaching style should complement students' learning preferences and needs (Lawter et al., 2014). Yet, there is a new stream of research that argues that deep learning is activated when students are engaged in unusual/unconventional learning approaches/types. In this sense, students are recommended to engage in learning approaches that are used less frequently and are, in a way, out of their comfort zone (Kolb & Kolb, 2009). The inclusion of a technology component is also argued to be essential for ensuring course success. Higher education students are increasingly interested in the technology component. Yet, there are ongoing debates on when and where technology could enhance learning outcomes. A lot of research is still required to help determine success factors in "blended" or "hybrid" education experiences (Abzug, 2015).

Business ethics education is said to face some possible challenges and potential obstacles that might jeopardize its effectiveness. From a student perspective, one of the main challenges of business ethics education is the perception that classes are "preachy." Ethics education might provoke negative emotions among students, including distress about how others would judge them, experiences of self-doubt, and guilt or shame feelings for one's previous or current ethical misconduct. Another challenge of ethics education for students is the issue of "relevancy," where students perceive ethics courses to follow an abstract and theoretical approach rather than a practical application or relevance (Simola, 2010). Despite that, such theoretical and systematic approaches might help in ethical decision-making, yet personal judgments and individual perspectives are considered to be the main predictors of one's actions and decisions (Mattison, 2000). Ethics is sometimes even argued to be hardly taught given that personal moral systems are based on a rigid set of values and beliefs that shape one's identity and can barely be reformed (Floyd et al., 2013). One more critical challenge of business ethics education is that students might have low involvement with the ethics courses due to perceptions about the unimportance of studying ethics courses on ethical business practices. With academic cheating being prevalent among business students, research advocates that the engagement in daily unethical behaviors (e.g., academic cheating) is regarded to be the main reason for more serious unethical conduct in the workplace. As for the challenges of business ethics education from a faculty perspective, teaching business ethics might pose a challenge for teachers as they question their expertise and proficiency in ethics. Teachers believe that the lack of "expert role" that has the necessary know-how about the rightful answers might endanger their integrity and influence on students. An additional challenge for teaching business ethics relates to teachers' concerns and worries about the likelihood of accidentally attacking or disputing students' held beliefs and values during the provision of guidelines on ethical behavior (Simola, 2010).

Business ethics education: Middle Eastern highlights

Questions about the role of institutions of higher education globally in preparing students for business ethics have always been raised and are becoming more pressing. This has been attributed to general moral decline in society (Jackling et al., 2007). The mission of higher education institutions is to advance knowledge to new generations by teaching and conducting research in order to equip students to be productive members of the community and the world (Choueiri & Myntti, 2012). Thus, higher education institutions exhibit a direct impact on society through alumni daily behaviors and knowledge. In such universities, faculty members' duties are not only limited to teaching courses in their area of expertise, but they are also entitled to discuss ethical issues related to their subjects (Tabsh et al., 2012). Scholars have argued that it is the responsibility of educators to contribute to the ethical knowledge of students because graduates study different domains of business, such as accounting, finance, and marketing, without understanding the ethical principles of each practice (Jackling et al., 2007; Seoudi & El-Bassiouny, 2010).

There are two main approaches for business ethics instruction in universities. One of them is through offering a full course in business ethics and the other is through integrating business ethics in class discussion, case studies, and coursework (Seoudi & El-Bassiouny, 2010). The business ethics courses include several domains. Corporate social responsibility (CSR) and sustainability are the main areas that higher education institutions in the Middle East are exhibiting.

CSR is essentially about the responsibility of business toward society (Al-Abdin et al., 2018; Jamali & Neville, 2011). CSR represents a framework for companies to enhance the well-being of society; it also acts as a competitive edge and improves the company image (Sherif, 2015). According to Carroll (1979), CSR includes four disciplines: economic, legal, ethical, and philanthropic. CSR in developing countries and the Middle East is recognized to have some different features than developed countries. For example, recent research indicated that small and medium enterprises in developing countries showed a high level of religious motives, which reflect natural altruistic philanthropic activities. Therefore, the cultural/religious values inherited in the Middle East is the primary distinction between CSR in developing and developed countries (Al-Abdin et al., 2018; Jamali & Neville, 2011).

Middle Eastern countries are recognized as a rapidly growing Muslim consumer segment with a strong influence and high purchasing power. There is widespread interest in considering Islamic marketing as a distinct domain, due to the religious values Muslims hold. Muslim countries have their own guiding system, based on Islamic law (*shar'iah*), which is adopted from the Holy Qur'an and the Sunnah. This means that, when faced with any ethical dilemma, Muslims are required to always choose what is regarded as *halal* (permissible) and refrain from what is *haram* (prohibited) according to the religious laws of *shar'iah* (Almoharby, 2011; Turnbull et al., 2016). In Islam, ethical conduct is framed by God's principles (Mohammad & Quoquab, 2016). For Muslims, ethical business conduct can only be achieved through true faith in *Allah* (God) (Almoharby, 2011). Subsequent to these principles, Muslims should be able to differentiate good deeds from bad deeds, moral actions from immoral ones, and moderate views from extreme ones (Mohammad & Quoquab, 2016). Teachings of the Qu'ran and Sunnah offer clear guidance to what is considered ethical or not in business dealings (Almoharby, 2011).

In most Middle Eastern countries, there is a schooling system in which curricula support Islamic values and ideologies. Such Islamic ideologies and ethics in Middle Eastern countries are expected to be different than those in Western societies (Turnbull et al., 2016). A Muslim businessman should not be involved in a business for selling alcoholic beverages or drugs, or invest in casinos, as they are regarded as *haram* in *shar'iah* (Abuznaid, 2009). *Riba* (unfair interest rate that is to be paid back on top of a loan) is strictly prohibited in Islamic *shar'iah* (Mohammad & Quoquab, 2016). Another example is related to socially responsible behavior being an obligatory aspect for businesses. Businessmen are required to give out a yearly percentage of their income (*zakat*) to the poor in their country (Abuznaid, 2009). Scholars have agreed on the importance of understanding Islamic ethics (Turnbull et al., 2016). Islam is considered as a comprehensive guide which creates distinct cultures that help in all aspects of life (Mohammad and Quoquab, 2016). In conclusion, "ethics, in a Muslim context have even been considered the nutrition of the soul" (Shahata, 1999, as cited in Turnbull et al., 2016). Therefore, business ethics education is naturally expected to be embedded in the higher education system.

"Islamicization" is a new approach that Muslim Middle Eastern countries are moving to as a way of reflecting their identities, values, and ethics and reducing negative materialistic societies. This is practiced in many Middle Eastern countries such as Jordan, Bahrain, Kuwait, and Oman through integrating Islamic values and ethics in their educational, social, and financial systems. This is done by creating prayer areas at the workplace and allowing or enforcing breaks during prayer times. Company policies are also changed to include a dress code for employees that fits more with Islamic laws. In terms of marketing activities in these countries, it is not allowed to publicly advertise products regarded as *haram*, such as alcoholic beverages, cigarettes, and drugs. Under the financial system, Islamic banks perform financial transactions that do not include interest on loans, and do not invest in any business that would be considered *haram* (Mohammad & Quoquab, 2016). With regards to education, a research study conducted in the United Arab Emirates (UAE) found that university students perceive ethics education and Islamic values as important guidance needed for their own ethical behavior. Thus, it is recommended to include Islamic ethics in the syllabus of business ethics courses to enhance student ethical behavior in the classroom and in their personal dealings (Fantazy et al., 2014).

CSR is increasingly playing a part in institutions of higher education curricula. CSR curricula are led by the Principles for Responsible Management Education (PRME). The PRME was launched in 2007 and is the first program to link between the United Nations and academic institutions, business schools, and universities. Examples of universities in the Middle East participating in PRME are Ain Shams University in Egypt, British and German universities in Egypt, the American University of Sharjah in the UAE, and the British University of Dubai. The PRME is considered as a global network for corporate sustainability and CSR. Moreover, PRME encourages the incorporation of universal values in teaching and research, which offers high-quality education. Universities in the world and the Middle East are applying the PRME principles. Participation in the PRME requires regular reporting and sharing of information among stakeholders on a database which facilitates the exchange of knowledge. Currently, universities are not only focusing on CSR education in the classroom setting, but they are also using CSR as part of their own competitive advantage (Sherif, 2015).

Sustainability and sustainable development is another form of teaching business ethics in higher education institutions. Sustainability is concerned with meeting the needs of the current generation without compromising the needs of future ones. Other scholars view sustainable development as a way to improve environmental problems (Al-Naqbi & Alshannag, 2018). The concept of sustainability has been a major focus of the world and has been mentioned in the United Nations' Millennium Development Goals. During the last decade, many universities have restructured their teaching methods, research, and operations to address sustainability challenges. The main objective of such sustainability initiatives is to improve students' sustainable actions. The best-known practice of sustainability education includes integrating environmental and sustainability issues in courses and research. This requires sustainability application in every research project and thesis. On the universities' operational level, they should manage energy, reduce waste, and review costing strategies to meet sustainability objectives (AlShuwaikhat et al., 2016).

The Association for the Advancement of Sustainability in Higher Education (AASHE) is an institution that supports sustainability in higher education through creating a shared community that promotes inclusive sustainability. According to the AASHE, sustainability comprises human, social, health, environment, and a better future for the coming generations. The activities of the AASHE aim to make sustainable development the norm among universities, facilitate the integration of sustainability in teaching and research, and enhance collaborations (Sherif, 2015). In order to help universities achieve such goals, the AASHE developed a self-reporting survey known as the Sustainability Tracking, Assessment & Rating System™ (STARS). This measurement tool is a transparent instrument to measure universities' sustainability performance (AlShuwaikhat et al., 2016; Sherif, 2015).

CSR and sustainability education in the Middle East are being recognized as critical content in business schools' course curricula, especially in Egypt, United Arab Emirates (UAE), and Lebanon. Egypt is expected to be the CSR hub in the Middle East. Cairo University introduced two courses in CSR and Social Entrepreneurship. Later, Suez Canal University and Port Said University, in addition to other private universities such as the German University in Cairo, Heliopolis University, and Nile University, followed the launch of such courses. Egypt is also working with international universities seeking to develop CSR. Examples of collaborations include Cairo University with the George Washington University and Notre Dame University (NDU) Sustainability Project 2012–2015 (Sherif, 2015).

UAE is another country in the Middle East leading CSR. The University of Dubai (UD) encourages entrepreneurship, and the campus itself is considered as a green construction. The UD was the first university to sign the PRME in the Middle East in 2008. As agreed in the PRME agreement, UD incorporated CSR into all the courses in Business Administration and students are obliged to take a "business and society" course (Sherif, 2015). Moreover, the UAE University has recently launched two research centers focusing on sustainable development, namely: National Water Center and the Emirate Center for Energy and Environment Research (Al-Naqbi & Alshannag, 2018).

The American University in Beirut (AUB) is also positioned as one of the universities in the Middle East following a systematic approach in retaining research in CSR. The AUB has pioneered publishing several research papers about CSR initiatives and challenges in the region. In addition to research, a Lebanese center for societal research has been developed in cooperation with Notre Dame University to enhance water, energy, and environmental development. The output of the center and research is applied in the classroom to engage students in implementing CSR and making better future decisions in the real world (Sherif, 2015).

The case of the Business Ethics course at the German University in Cairo, Egypt

Course description

Modern management scholarship highlights the importance of contemporary issues that impact individual and collective corporate performance and decision-making. This is reflected by globally renowned bodies including the Social Issues in Management (SIM) division of the Academy of Management, the Giving Voice to Values (GVV) program at the University of Virginia Darden School of Business, US, the International Association for Business and Society (IABS), and many others. Amongst the most important contemporary issues in modern times, business ethics emerged as a global imperative, given the many criticisms of the management function and its impact on society. In Egypt, the emphasis on sustainability and ethics is growing with the parallel growth of corporations introducing corporate social responsibility (CSR) departments. The discussion is also relevant to the mass-combating of corruption that occurred after the 25 January 2011 revolution, when civil society organizations called for fighting corruption on all levels in the country.

In an intellectually stimulating environment, the aim of this course is to expose students to issues of sustainability, ethics, and corporate social responsibility. The course will also equip students with hands-on international tools that they can effectively use in decision-making relevant to day-to-day ethical dilemmas that they face in work settings. However, to overcome the drawbacks of traditional ethics case study approaches that are focused on ethical dilemmas, the course utilizes Giving Voice to Values cases (as described below) that are also specific to the Egyptian context.

The course, therefore, presents a novel methodology for decision-making, namely the Giving Voice to Values (GVV) methodology, which will be described in detail in the coming section of this chapter. In general, the course is intended to acquaint students with state-of-the-art knowledge of business ethics, corporate social responsibility (CSR), and sustainability in terms of both theory and application. The course provides a closer look into the stakeholders involved in a company's business relationships. Throughout the course material, we touch upon several different topics that broaden students' perspectives on relevant best practices, including codes of ethics development, CSR, international cases of unethical practice, factors affecting ethical decision-making, and sustainability.

Intended learning outcomes

Table 3.1 ILOs of business ethics course

Knowledge and understanding
• To develop a profound knowledge of global sustainability and social responsibility concerns as related to the interface between business and society. • To understand the basic concepts of corporate social responsibility as an important contemporary topic. • To comprehend the idea of a code of ethics and the different ways to develop it. • To understand the value of functional ethics. • To develop profound knowledge in new areas related to social responsibility such as social marketing. • To learn the importance of group work as reflected in course group case and article presentations, and an empirical/applied project.

Intellectual skills

- To be able to reason through different types of ethical dilemmas using the GVV methodology.
- To develop a further analytical understanding of the difficulties involved in ethical choices and their implications on business strategy.
- To be able to balance out the necessary requirements of developing an industry and corporate code of ethics, especially with relevance to the business environment in Egypt.
- To develop the student's ability to apply course knowledge and analyze complex situations through selected cases and articles.

Professional and practical skills

- To learn about the value of business ethics to today's organizations at large.
- To improve the student's ability to critically and conceptually analyze challenging GVV cases.
- To provide the student with the necessary skills to audit corporate ethics.
- To develop the student's teamwork and team management capabilities.
- To master the different career opportunities in corporate social responsibility.

General and transferrable skills

- To improve the student's ability to critically and conceptually analyze challenging GVV cases.
- To provide the students with the necessary skills to audit corporate ethics.
- To improve students' report preparation and data presentation skills.
- To develop student teamwork and team management capabilities.

Referencing and methods

In an attempt to mitigate the challenges of business ethics education mentioned in international literature and highlighted previously, the course focuses on experiential learning approaches that also include pedagogies that tackle the affective component to decision-making, including case studies, guest speakers, and world-class movies.

The course is, therefore, based on a multiplicity of learning methodologies aimed at enhancing students' learning experience. The course lectures are based on Trevino and Nelson (2011). A supplementary text is Ferrell et al. (2008).

The Giving Voice to Values case studies are available online at www.givingvoicetovalues.org.

The project is based on developing a code of ethics for an organization or function of choice. A reference text for the project Collins (2009).

Moving toward the Giving Voice to Values (GVV) methodology

The Giving Voice to Values (GVV) methodology was developed by Professor Mary Gentile in 2009, with the Aspen Institute as incubator and the Yale School of Management as founding partner. From 2009–16, the initiative was based at Babson College; it is now based at the University of Virginia Darden School of Business. The GVV methodology gives guidance into how leadership and values-driven decision-making in actual business practices and in business education could be conducted. The GVV curriculum offers a variety of business cases from different domains, most of which are free for faculty members to download (Gentile, 2012; Giving Voice to Values, 2018). Cases revolve around business problems that offer an open discussion for practitioners and educators about the possible ethical decisions that can be taken to solve the business dilemma at hand. The protagonist of the different cases is mostly an employee who is pressured into unethical practices by their manager or boss. This raises the discussion questions about how one should act on one's values under such pressure. Such discussions give ways to analyze how to strategize and act out solutions that voice an employee's values (Giving Voice to Values, 2018).

The GVV curriculum differs from other ethics-focused courses, where the usual practice is to present a managerial ethical dilemma and ask the audience about the best decision to take and whether this would be considered ethical or not (Arce & Gentile, 2015; Gentile, 2010; Giving Voice to Values, 2018). Under GVV, managers or protagonists of a certain dilemma know that they need to act on their value, know what is ethical and what is not, but need guidance on how to act most effectively (Gentile, 2010; Giving Voice to Values, 2018). This approach is different from other business ethics courses, as it does not put the ethical business dilemma into a mere debatable question, but rather focuses on how to apply what is already regarded as right. Those business cases ask the question of what the right/ethical course of conduct is. The protagonist of the case does not know what the ethical solution to the business dilemma might be. Under GVV, the right/ethical decision is already known. It closes the door in the face of those who would knowingly choose to act in an unethical manner. The question under GVV is rather how to implement the right/ethical decision. Thus, the GVV approach is a "post-decision-making approach." This means that the GVV approach shows people that there is always a way to act on one's values, even if pressured otherwise (Arce & Gentile, 2015).

The GVV methodology is based on values. Many values are widely agreed upon and shared among people from different countries and cultures. This is an important starting point, as it makes the idea of voicing one's values in an organization easier. Knowing that people share common values, even when they have disagreement about a course of action at work, makes voicing and acting on values more manageable (Gentile, 2010). In cases where some values might differ between employees and managers, due to different cultures and religious backgrounds, GVV offers employees the tools and skills to correctly strategize the course of action needed to react. Many Muslims, for instance, are employed in Western countries. These individuals might be faced with situations at work, where managers require a task that is regarded as ethical by the norms and culture of the country, yet oppose those of the

Muslim employee. For instance, an advertising agency might want to execute an ad displaying nudity. Although this is accepted in many Western cultures, it is prohibited for Muslims to be involved in such projects. Rewards coming out of such projects would also be regarded as *haram*. In such instances, the Muslim employee knows that they should not conduct the task as it would be regarded as *haram* religiously, but would not know how to oppose or reject the task. This is where the GVV methodology comes into action, showing people that they should not be forced into acting against their values and ethical system.

The focus of regular business courses is usually divided into creating *awareness* of an ethical problem and starting an *analysis* about how to deal with it. GVV cases move away from a mere awareness of an ethical dilemma and its analysis, into actual *action* discussion (Arce and Gentile, 2015; Gentile, 2012). While discussing action and arguments for it, the course audience is asked to use the vocabulary of the business domain highlighted in the case in order to move it as close as possible to the real-life situation that this audience may encounter in their own business. Thus, they are requested to move away from philosophical answers towards practice-oriented ones (Arce & Gentile, 2015; Gentile, 2012). The case discussed is referred to in the GVV curriculum as Case A. After discussion, the audience is presented with Case B, which shows them the actual value-based action taken by the business in question for further discussion (Arce & Gentile, 2015).

The audience gets to evaluate cases by reflecting on their own behavior in a similar situation. When the audience realizes through discussion and self-reflection that they might not have acted on their values at all times, a sense of enlightenment and self-empowerment for improvement occurs. The GVV methodology hence focuses on strengthening the human will to stand up against unethical pressure and to respond to the voice of values within them. The practice offered through the different case analyses reduces the surprise element related to facing dilemmas arising in the actual workplace of an individual. Individuals become well trained on how to stand up to unethical pressure and how to effectively give voice to their values and put them into action (Gentile, 2010). This is referred to as "moral muscle memory," as individuals are being trained to think strategically and efficiently to solve any real-life problems they face. Thus, GVV offers everyone in business a chance to be an active player in decision-making. It also helps practitioners to shift their focus away from the mere profit-oriented one to one that questions the purpose of the business before making any decision (Arce & Gentile, 2015).

Cases under GVV focus on the post-decision-making step (Arce & Gentile, 2015; Gentile, 2010). This is the case because the protagonist of a case and the audience analyzing the case already know the decision that needs to be taken but ask how it should be done in an effective manner. A GVV case asks its audience three main questions (Arce & Gentile, 2015): (1) Who are the involved parties in this situation and what would they gain or lose?

(2) What action needs to be done and what is the logic? (3) How can other parties against this action be convinced of it? Groups working on different cases can be asked to answer these three questions using several scenarios and arguments to offer more ground for comments and discussion.

The focus of the GVV methodology is not to change the worst in its audience but rather to empower the good in them. This is why there is high emphasis on showing people how to strategize and effectively work towards success while being their best throughout the process. Hence, the focus of the GVV approach is on *pragmatists* rather than trying to change the *opportunist* or to work with *idealists* (Arce & Gentile, 2015; Gentile, 2012).

Gentile (2012) highlights eight aspects that make the GVV methodology different from other business ethics course approaches. The first difference is the *question asked*. GVV does not ask about what the right thing is, but asks how to implement the already established right approach. The second difference is the *focus of the problem*. GVV cases do not focus on *gray* areas where people will always reach disagreement about what to do, but rather take clear-cut issues where the right decision is clear and focus on how managers and decision-makers can get the "right thing" done. Accordingly, the focus is not on telling the audience what not to do but rather how to do what is right and take action towards it. The third difference is referred to as the *thought experiment*. The audience is given a list of assumptions regarding the value system and is asked to regard them as true in order to develop feasible action plans for the problem at hand. Room is left for discussing these assumptions, but the audience is later asked to use them equally for the case analyses and action plan development. The fourth difference is related to focusing on *rehearsal*. Members act out the scripts they have written for action plans in front of other audience members in order to receive feedback that could take place in reality between them and their managers and customers. The fifth difference is related to the *structure of the case studies*. GVV cases are not long and they do not always show top managers as decision-makers. Instead, employees of different levels are shown and they know the right thing to do but do not know how to take action. GVV cases are mostly based on scenarios where the protagonist was able to act on their values in reality. The sixth difference is that GVV cases focus on *prescription and implementation*. This means that the audience spends most of its time scripting how to strategize and implement an action in a manner allowing the focus on values rather than spending most of the time analyzing what should be done. The seventh difference is about the *use of research*. The audience is given a chance to think about and research how to improve their own behavioral biases and to use research in that area, rather than going with the established research about general expected biases by others. The eighth and last difference is in *flipping role play*. Instead of making one member act as an ethical member against unethical opposition, all work as coaches or trainers on how to best implement the action plan developed.

Conclusions

Ethics education has emerged as an important research domain that has practical implications for the conduct of day-to-day business. The question remains, however, how ethics can be taught. One of the prominent emerging methodologies for teaching ethics is the Giving Voice to Values (GVV) pedagogy that centers around the premise that the decision-maker knows the right course of action to take and is searching for ways of implementation. This chapter highlighted the case of the Business Ethics course at the German University in Cairo (GUC) that has transitioned from the use of ethical dilemmas in case studies to the GVV methodology, where the methodology was highlighted in detail. Utilizing the GVV methodology enhances the benefits of the course to students. As the business ethics course is a standalone course, it had the potential to go beyond the usual approach of teaching ethics to focusing on providing students with the skills and tools to survive in the face of actual ethical dilemmas they might face in the workplace. Although case-specific problems are discussed, students come out of the course with general strategies that can be used in practice. Several cases used at GUC are about ethical dilemmas that were faced by managers and employees in Egypt. Using cases from Egypt makes the class discussion more relevant, as students can relate more to the protagonist, the culture, and context of the problem. Cases used are from different industries, such as the pharmaceutical and healthcare industry (Hamed et al., 2015), the financial and banking sector (Salem & El-Bassiouny, 2015), and citizenship and volunteerism movements (Darrag & El-Bassiouny, 2015). This course experience can be replicated in other universities, not only in the Middle East and North African (MENA) region, but also on a global level.

Note

1 For more information on the GVV pedagogy, please visit www.GivingVoiceToValues.org and www.MaryGentile.com/.

References

Abuznaid, S. (2009). Business ethics in Islam: the glaring gap in practice. *International Journal of Islamic and Middle Eastern Finance and Management*, 2(4), 278–288.

Abzug, R. (2015). Predicting success in the undergraduate hybrid business ethics class: conscientiousness directly measured. *Journal of Applied Research in Higher Education*, 7(2), 400–411.

Al-Abdin, A., Roy, T., & Nicholson, J. D. (2018). Researching corporate social responsibility in the Middle East: the current state and future directions. *Corporate Social Responsibility and Environmental Management*, 25(1), 47–65.

Al-Naqbi, A. K., & Alshannag, Q. (2018). The status of education for sustainable development and sustainability knowledge, attitudes, and behaviors of UAE university students. *International Journal of Sustainability in Higher Education*, 19, 566–588.

Almoharby, D. (2011). The current world business meltdown: Islamic religion as a regulator. *Humanomics*, 27(2), 97–108.

Alshuwaikhat, H., Adenle, Y., & Saghir, B. (2016). Sustainability assessment of higher education institutions in Saudi Arabia. *Sustainability*, 8, 750.

Arce, D. G., & Gentile, M. C. (2015). Giving Voice to Values as a leverage point in business ethics education. *Journal of Business Ethics*, 131(3), 535–542.

Brook, C., & Christy, G. (2013). Doing right in business: can action learning develop moral sensitivity and promote ethical behaviour?. *Action Learning: Research and Practice*, 10(3), 214–229.

Carroll, A. B. (1979). A three-dimensional conceptual model of corporate performance. *Academy of Management Review*, 4(4), 497–505.

Chavan, M., & Carter, L. M. (2018). Sustainable business ethics education. In *Meeting Expectations in Management Education* (pp. 149–170). Palgrave Macmillan, Cham.

Choueiri, T., & Myntti, C. (2012). The AUB neighborhood initiative: social responsibility in a university's backyard. In *CSR in the Middle East* (pp. 158–175). Palgrave Macmillan, London.

Christensen, L. J., Peirce, E., Hartman, L. P., Hoffman, W. M., & Carrier, J. (2007). Ethics, CSR, and sustainability education in the Financial Times top 50 global business schools: baseline data and future research directions. *Journal of Business Ethics*, 73(4), 347–368.

Collins, D. (2009). *Essentials of Business Ethics: Creating an organization of high integrity and superior performance*. Wiley & Sons, Inc.

Craft, J. L. (2013). A review of the empirical ethical decision-making literature: 2004–2011. *Journal of Business Ethics*, 117(2), 221–259.

Darrag, M., & El-Bassiouny, N. (2015). 'Sharek' and employee volunteerism: A social enterprise succeeds in the corporate world. In *Giving Voice to Values Curriculum Collection*. Darden Business Publishing, University of Virginia.

Downe, J., Cowell, R., & Morgan, K. (2016). What determines ethical behavior in public organizations: is it rules of leadership? *Public Administration Review*, 76(6), 898–909.

Fantazy, K., Abdul Rahim A., & Al Athmay, A. A. (2014). Ethics and religion in higher education: evidence from United Arab Emirates universities. *International Journal of Commerce and Management*, 24(2), 180–196.

Ferrell, O.C., Fraedrich, J., & Ferrell, L. (2008). *Business Ethics: Ethical Decision making and cases*. Houghton Mifflin Company, Boston.

Floyd, L. A., Xu, F., Atkins, R., & Caldwell, C. (2013). Ethical outcomes and business ethics: toward improving business ethics education. *Journal of Business Ethics*, 117(4), 753–776.

Gentile, M. C. (2010). *Giving Voice to Values: How to speak your mind when you know what's right*. Yale University Press, New Haven, CT.

Gentile, M. (2012). Values-driven leadership development: where we have been and where we could go. *Organization Management Journal*, 9(3), 188–196.

Giving Voice to Values. (2018). Retrieved from www.darden.virginia.edu/ibis/initiatives/giving-voice-to-values/. Accessed on 2 June 2018.

Hamed, S., Shaaban, Y., Abutaleb, S., Nashaat, D., Alaa, S., & Asar, A. (2015). Good health or wealth: can they co-exist? In *Giving Voice to Values Curriculum Collection*, Darden Business Publishing, University of Virginia.

Jackling, B., Cooper, B. J., Leung, P., & Dellaportas, S. (2007). Professional accounting bodies' perceptions of ethical issues, causes of ethical failure and ethics education. *Managerial Auditing Journal*, 22(9), 928–944.

Jamali, D., & Neville, B. (2011). Convergence versus divergence of CSR in developing countries: an embedded multi-layered institutional lens. *Journal of Business Ethics*, 102(4), 599–621.

Jones, T. M. (1991). Ethical decision making by individuals in organizations: an issue-contingent model. *Academy of Management Review*, 16(2), 366–395.

Kerr, S. (2014, November 24). CEO suspended at Saudi telecoms group hit by accounting scandal, *Financial Times*, Retrieved from www.ft.com/content/c186cf94-69cc-11e4-8f4f-00144feabdc0.

Kolb, A. Y., & Kolb, D. A. (2009). The learning way: meta-cognitive aspects of experiential learning. *Simulation & Gaming*, 40(3), 297–327.

Lam, C. F. (2004). Understanding the ethical decisions and behaviours of Hong Kong business managers: an implication for business ethics education. *Management Research News*, 27(10), 69–77.

Lawter, L., Rua, T., & Guo, C. (2014). The interaction between learning styles, ethics education, and ethical climate. *Journal of Management Development*, 33(6), 580–593.

Mattison, M. (2000). Ethical decision making: the person in the process. *Social Work*, 45(3), 201–212.

Mintz, S. (1996). Aristotelian virtue and business ethics education. *Journal of Business Ethics*, 15, 827–838.

Mohammad, J., & Quoquab, F. (2016). Furthering the thought on Islamic work ethic: how does it differ? *Journal of Islamic Marketing*, 7(3), 355–375.

Neill, M. S. (2017). Ethics education in public relations: differences between stand-alone ethics courses and an integrated approach. *Journal of Media Ethics*, 32(2), 118–131.

Rest, J. R. (1986). *Moral Development: Advances in research and theory*. Praeger, New York.

Rest, J. R. (1994). Background: Theory and research. *Moral Development in the Professions: Psychology and Applied Ethics*, 1–26.

Salem, R., & El-Bassiouny, N. (2015). AAIB addresses social and environmental risks in loan assessment. In *Giving Voice to Values Curriculum Collection*, Darden Business Publishing, University of Virginia.

Seoudi, S., & El-Bassiouny, N. (2010). Egyptian business students' perceptions of ethics: the effectiveness of a formal course in business ethics. *Journal of Business Leadership*, Fall, 24–48.

Shahata, H. (1999), *Business Ethics in Islam*. Al-Falah Foundation for Translation, Publishing and Distribution, Cairo.

Sherif, S. F. (2015). The role of higher education institutions in propagating corporate social responsibility case study: universities in the Middle East. *International Journal of Education and Research*, 3(1), 217–226.

Simola, S. (2010). Use of a "coping-modeling, problem-solving" program in business ethics education. *Journal of Business Ethics*, 96(3), 383–401.

Simola, S. (2014a). Facilitating embodied learning in business ethics education: the use of relational sculpting. *Journal of Applied Research in Higher Education*, 6(1), 75–97.

Simola, S. (2014b). Teaching corporate crisis management through business ethics education. *European Journal of Training and Development*, 38(5), 483–503.

Tabsh, S. W., El Kadi, H. A., & Abdelfatah, A. S. (2012). Faculty response to ethical issues at an American university in the Middle East. *Quality Assurance in Education*, 20(4), 319–340.

Trevino, L. (1992). Moral reasoning and business ethics: implications for research, education, and management. *Journal of Business Ethics*, 11, 445–459.

Trevino, L., & Nelson, K. (2011). *Managing Business Ethics: Straight talk about how to do it right.* John Wiley & Sons, Inc., USA.

Turnbull, S., Howe-Walsh, L., & Boulanouar, A. (2016). The advertising standardisation debate revisited: implications of Islamic ethics on standardisation/localisation of advertising in Middle East Islamic States. *Journal of Islamic Marketing*, 7(1), 2–14.

Wang, L., & Calvano, L. (2015). Is business ethics education effective? An analysis of gender, personal ethical perspectives, and moral judgment. *Journal of Business Ethics*, 126, 591–602.

Welch, J. (2019). The Volkswagen recovery leaving scandal in the dust. *Journal of Business Strategy*. Available at: https://doi.org/10.1108/JBS-04-2018-0068.

4 Education for sustainable development

A means for infusing social responsibility in higher education in Egypt

Heba El-Deghaidy

Introduction

There is strong evidence to suggest that sustainable development practices in the Arab region, including Egypt, are lagging behind (EL-Deghaidy, 2012; Othman & EL-Deghaidy, 2008; UNESCO, 2008). For the region to catch up, various goals and agendas were signed that help put these countries on track. Among these agendas are the Millennium Development Goals (MDGs) and the United Nations Educational, Scientific and Cultural Organization (UNESCO) Decade for Sustainable Development (DESD) 2005–2014 (UNESCO, 2005; UNESCO, 2014a). The MDGs consist of eight goals where each has measurable targets and deadlines aiming to improve the lives of the world's poorest people. The leaders of 189 countries including Arab countries signed this declaration during their meeting at the United Nations Millennium Summit in 2000. The declaration set the year 2015 to achieve various goals, from providing universal primary education to avoiding child and maternal mortality (UN, 2011). As for the DESD, it aimed to ensure that the ecological and social issues facing various nations are resolved through: promoting and improving quality education; reorienting educational programs; building public understanding and awareness; and providing practical training (UN, 2005). Although education is the first step towards a successful future and nurturing the human mind to foster a generation renowned for creativity, innovation, leadership, and achievement, educational systems in Arab countries are facing numerous challenges themselves and this decreases their capacity to achieve the desired goals of DESD (UNESCO, 2008). As a follow-up of the international efforts to improve quality of life, the Rio + 20 conference in 2012 highlighted its concern in its search for solutions and suggestions for a future where aspects related to environmental, social, economic, and cultural considerations are balanced. Unfortunately, there are considerable consequences resulting from humanity's constant unsustainable practices. These acts include the rise of carbon emissions and high consumption rates, which have an impact on future generations, who will be facing dramatic challenges to resolve some of the actions this current generation is responsible for. Among these challenges are poverty, climate change, global warming, inequality, and various other issues, all of major concern (World Economic and Social Survey, 2013).

After reviewing the achievements of the DESD and MDGs and finding that more needs to be done to continue and consolidate their achievements, the United Nations developed the 17 Sustainable Development Goals (SDGs) in 2016 to be achieved by developing and developed countries alike. The SDGs include a vision of building systematic partnerships with the private sector. The 17 SDGs can be divided into three categories. The first seven goals can be viewed as an extension of the MDGs, while the second group (goals 8–10) focus upon inclusiveness (jobs, infrastructure, industrialization, and distribution). The third group of goals (11–17) focus upon issues related to sustainability and urbanization (Kumar et al., 2016).

> Since the focus of this chapter is on Egypt, it is worth highlighting that Egypt adopted an ambitious approach that indicates its firm commitment and dynamic innovation towards tackling issues related to sustainable development (SD). This is through a transformative agenda aligned with the 2030 Agenda articulated by the SDGs (Ministry of Planning, Monitoring and Administrative Reform, 2018; UN, n.d.). The transformative agenda is expressed by Egypt's 2030 strategy for sustainable development, Egypt's "Vision 2030." The vision spans over three dimensions of sustainable development – economic, social, and environmental dimensions – and outlines the broader principles which will guide Egypt in pursuing its developmental goals. The main aim of Vision 2030 is for Egypt to:
>
> possess a competitive, balanced and diversified economy, dependent on innovation and knowledge, based on justice, social integrity and participation, in a balanced and diversified ecological collaboration system, investing the ingenuity of place and human capital to achieve sustainable development and to improve Egyptians' quality of life, in a state-driven process, with the full participation of all relevant stakeholders.
>
> (Ministry of International Cooperation, 2016, p. 6)

Evidence of Egypt's commitment to the SDGs is expressed in its involvement in capacity-building and technical support partnerships via the Egyptian Agency of Partnership for Development (EAPD), Participation in the COP21 climate change conference in Paris, and through submitting its report on the Intended Nationally Determined Contributions (INDCs) towards achieving the objectives of the United Nations Framework Convention on Climate Change (UNFCCC) (Ministry of International Cooperation, 2016). Despite such commitments, there are areas where Egypt is still struggling. These include difficulty in closing the gender gap between boys and girls and expanding access to basic education (Ministry of International Cooperation, 2016).

As indicated in the MDGs, DESD, SDGs, and Egypt's Vision 2030, there is great emphasis on the role of education. A report by Salem et al. (2018) denotes that there are five SDGs where Egypt is underperforming, one of which is "quality of education." It seems that the indicators are not sufficient to measure these goals and that further indicators are needed to establish their achievement (Salem et al., 2018). EL-Deghaidy (2012) also believes in education as the first step towards a successful future and core in human capital through education for sustainable development (ESD), which is distinct from environmental education yet complementary (McKeown & Hopkins, 2003). For investing in human capital this requires a combination of teaching strategies that allows for transformative learning in a course with an ESD philosophy and practice.

Discourse on ESD

Education for sustainable development has been conceptualized differently in terms of its content, pedagogy, and the competences and skills needed (Wals & Blaze Corcoran, 2006; Tilbury & Wortman, 2004). The main reason behind this variation is that SD by nature is contested (Landorf et al., 2008), due to the different perspectives and interests in the social issues it deals with. The Bruntdland Report laid out the framework for sustainability through the lens of environmental, economic, social, and political concerns (World Commission on Environment and Development [WCED], 1987), while the link between SD and education was established in chapter 36 of an action plan titled "Agenda 21" during the 1992 UN Conference on Environment and Development. Wals and Kieft (2010) argue that education is by no means a value-free process, as there are discussions about whether education should be concerned with "social reproduction" or about "social transformation." There are key differences between these two roles, as the former assumes that learners are "to accept their role within society and the workforce. They are obedient, deferential, and compliant as they take their place within hierarchical and authoritative social structures and power relationships" (p. 8). The latter has a contrasting view as it perceives the learner as an active member involved in the democratic process of the community that they are part of. Since the role of learners differs, their role in the society and context differ as well. Vare and Scott (2007) outline two inter-related and complementary approaches to ESD that reflect these differing roles, namely ESD 1 and ESD 2. ESD 1 is concerned with "raising awareness of the necessity for change and 'signposting' goods and services that will reduce the ecological footprint of our activities" (p. 193). As such, the focus of education is on guiding the learner to reduce the ecological impact through promoting informed and skilled behaviors. The approach assumes that positive environmental and social benefits can be obtained through a combination of incentives directed towards the learner. As such, it inherently perceives the role of education as instrumental through top–down initiatives which could be

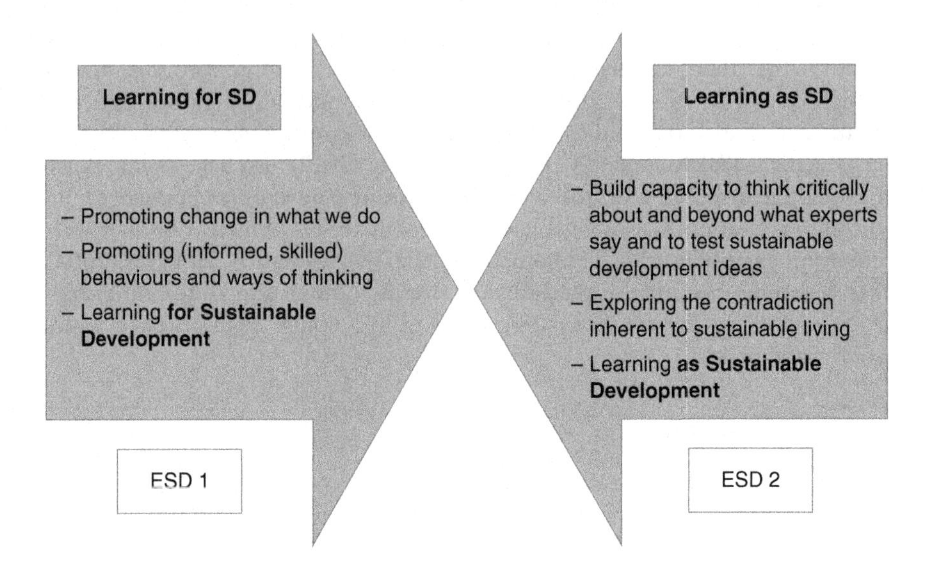

Figure 4.1 ESD 1 and ESD 2.
Source: Vare and Scott (2007).

useful to solving some simple issues. In ESD 2, there is a complete reframing of the role of education in regards to sustainable development, as it is about the "learning process" rather than the "outcome." Therefore, ESD 2 aims at creating capacity through a reflective approach as it reframes learning as a form of sustainable development, as the learner makes decisions and thinks critically and deeply beyond what experts say. Nonetheless, ESD 1 and ESD 2 are perceived as complementary approaches that are both needed.

Social responsibility (SR)

Social responsibility is one of the major attitudes that can enhance the cohesion and unity of any society. For this particular reason, programs and institutions, whether locally or globally, are working on developing the sense of social responsibility (Weiss, 2016). For clarity, social responsibility is defined as "one's sense of duty to the society in which he or she lives" (Brondani, 2012, p. 609). Scales et al. (2000) clarified the target group towards which one should have such a duty by providing a sense of caring for and helping the people who live in the community. When investigating the outcomes associated with learners' involvement in activities related to social responsibility in their societies, three major domains are indicated. These include the personal, academic, and social domains. The focus of this chapter is mainly on the academic and social domain. The academic domain links the service provided to society to the formal curriculum (Furco, 1994; Scales et al., 2000;

Wade, 1997). Students' involvement in such programs and activities can result in developing their academic achievement and skills, as they have a chance to apply academic concepts and practice abstract thinking (Conrad & Hedin, 1982; Newmann & Rutter, 1983). As for the social domain, there is agreement that the engagement in community service programs can have a positive impact on participants' social development. Civic engagement, social responsibility, and society's political organization and morale are key outcomes of the social domain (Brunelle, 2001; Conrad & Hedin, 1982; Raskoff & Sundeen, 1999; Wade, 1997).

Types of social responsibility

In portraying types and models of social responsibility, the following section presents three major concepts or types. These include corporate social responsibility (CSR), university social responsibility (USR), and finally individual social responsibility (ISR). CSR as a concept developed after the First World War as business leaders took responsibility to comply with certain practices (Windsor, 2001). Definitions of CSR range from business ethics, to sustainability, to corporate citizenship. Some companies simply see CSR as "the right thing to do"; while others see it as a strategic differentiator for their company and a means to achieving greater business value (Amin-Chaudhry, 2016). Maignan and Ferrell (2004) and Thomas and Nowak (2006) define CSR as simply a "social obligation." Others define it as the relationship of corporations with society as a whole, and the need for corporations to align their values with societal expectations to be engaged in social development (Temkar, n.d.) and to avoid conflicts and reap tangible benefits (Camilleri, 2017a) while achieving its balance through the "triple-bottom-line approach" (Alikihc, 2015). Contrary to the definitions above, Camilleri (2017b) stated that the concept of CSR is being challenged by those who want corporations to move beyond transparency, ethical behavior, and stakeholder engagement. Nonetheless, there are four different categories for CSR – ethical responsibility, philanthropic responsibility, environmental responsibility, and economic responsibility (see Figure 4.2) – or what could be combined under the "triple-bottom-line" (TBL) approach that achieves a balance of economic, environmental, and social imperatives, while at the same time addresses the expectations of shareholders and stakeholders (Bozkurt & Ergen, 2015). The TBL is also an approach tied closely to financial reporting as it extends it further to performance reporting on sustainable development, such as the use of the Global Reporting Initiative (GRI) standards (GRI, n.d.). With this extension, the aim is to ensure efficient and developed corporate performance in terms of sustainability management (Elkington, 1997). The TBL approach aligns with the main roots of CSR that are concerned with environmental degradation (Smith, 1993) and extends it further to the current responsibilities of corporates to "not only avoid environmental harm but also to protect and improve the natural environment. The rights of future generations is another important dimension in stakeholder management" (p. 417), as its conceptual background for such responsibilities is linked to the notion of SD. For corporates to identify how well they are doing in terms of their CSR, scales can be used, such as those found in the literature (i.e., Aupperle, 1984; Hopkins, 2005; Maignan & Ferrell, 2000; Quazi & O'Brien, 2000; Turker, 2008; Wood & Jones, 1995).

Universities and higher education institutions are part of any society and can be perceived as entities similar in some aspects to corporates due to their social obligations and responsibilities. Universities have stakeholders to attend and satisfy (Sanchez-Hernandez & Mainardes, 2016) in a changing marketplace that applies customer-oriented principles to educational institutions (DeShields et al., 2005). Malone (2018) provided a vision on how United Nations University can be involved in promoting social responsibility. This is reflected in the following three functions: universities are perceived as interdisciplinary research institutions that focus on major global problems; a think-tank that translates the outcome of research into policy recommendations; and finally, universities are postgraduate training and capacity development organizations to build capacity. So, to transform universities into more sustainable institutions necessitates rethinking what universities are doing in four key areas: (1) curriculum, teaching, and learning; (2) research and development; (3) institutional/administrative operations; and (4) partnerships and outreach (Figure 4.2). These are driven by the three realms of sustainable development: (1) social and economic justice; (2) ecological integrity; and (3) the well-being of all living systems on the planet through an integrative and cross-cutting manner. Both integrative and cross-cutting approaches are

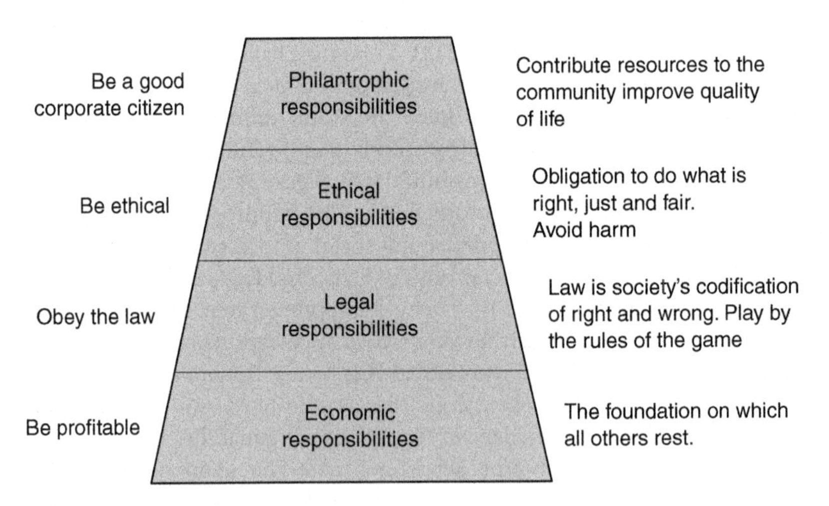

Figure 4.2 CSR pyramid.

Source: Alikihc, (2015, p. 99).

means used to introducing knowledge, skills, and dispositions in a meaningful manner that can help shape and form a sustainable world. For universities to monitor how well they are doing in terms of their university social responsibilities (USR), scales and tools such as those developed by Sammalisto and Arvidsson, (2005) and Setó-Pamies and Papaoikonomou (2016) are examples of researchers' interest to establish universities' roles towards society (Filho et al., 2019).

Infusing sustainability in higher education institutions

Higher education institutions (HEIs) are perceived to have a major role "tied so intricately to economic, social and environmental fabric of the modern world" (Wells, 2017, p. 31). The role of HEIs has changed over the years "from preservers of culturally revered forms of knowledge, through producers of skilled labour associated with a workforce-planning approach, to a

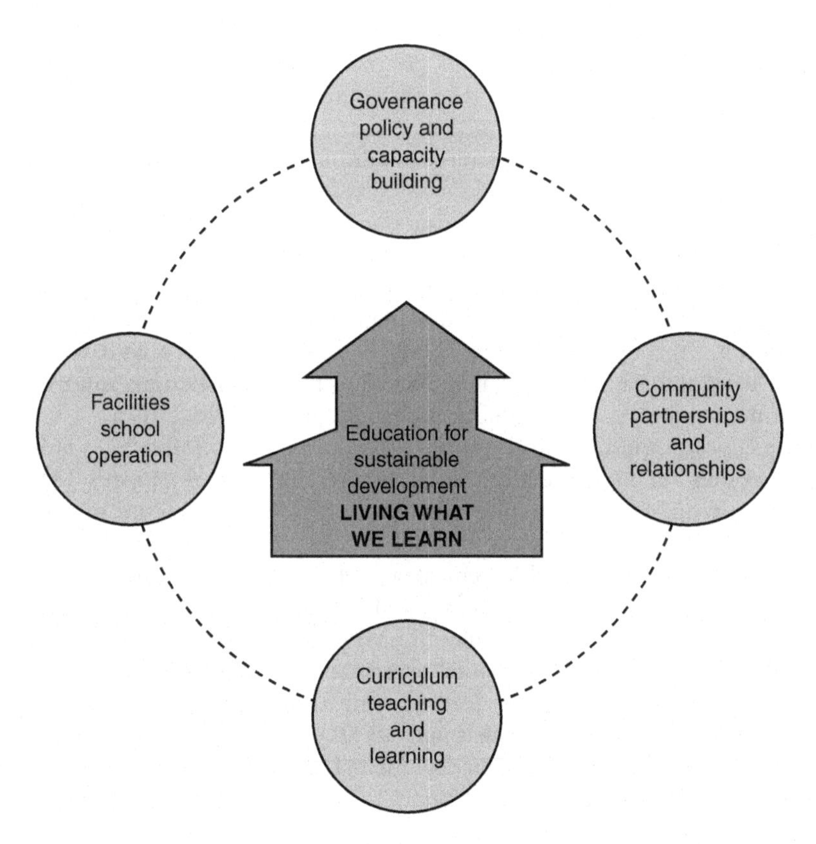

Figure 4.3 The "whole-institution approach."
Source: UNESCO (2014b, p. 89).

more recent perception as agents of social change and development" (Grau et al., 2017, p. 42). For universities to optimize their roles as agents of change, adopting a "whole-of-university" approach to sustainability is suggested (McMillin & Dyball, 2009). The approach is based on explicit linkage between research, education, and operational activities of any HEI seeking to apply a sustainable system (Gallardo-Vazquez & Sanchez-Hernandez, 2014). The Tallories declaration was a starting point for universities to commit to such an approach by ensuring its commitment to the 10-point action plan with 363 universities in 77 countries worldwide signing the declaration (Weiss, 2016). In 2017, Colombian universities (Latin America and the Caribbean) validated a proposal entailing indicators and management and reporting systems as part of universities' social responsibility for the Principles for Responsible Management Education (PRME) which help achieve the SDGs and the 2030 Agenda. PRME is open to the international community for universities to utilize these indicators and respond to the various challenges of the 21st century (Filho, 2017). PRME is considered to be the first organization that provides links between the United Nations and HEIs and provides a global network to promote sustainability, CSR, and a framework of commitment for universal values in programs and research among over 919 academic institutions in more than 90 countries. There are five major components in the PRME indicators, known as "areas." These are leadership and strategy, teaching, research, operational administration, and extension or social projection.

With the need to resolve complex issues and challenges locally and globally, as identified by the SDGs, HEIs are at the forefront to come up with innovative programs, creative and collaborative research agendas, and projects that require crossing geographical and disciplinary boundaries. Higher education institutions, through their faculty, facilities, and programs, could have an impact in developing human capital and shaping future generations towards empowerment and social responsibility as change agents (Wells, 2017).

Higher education institutions can utilize various means to infuse SD. This includes reorienting higher education programs and courses towards SD. This chapter focuses on illustrating the experience students at the undergraduate and graduate levels went through while enrolled in a course on education for sustainable development at a higher education institution. A previous example of courses being reoriented is that from a Trans-European Mobility Scheme for University Studies (TEMPUS)-funded project managed to include HEIs and non-governmental organizations (NGOs) in three Arab countries (Egypt, Lebanon, and Jordan) and three European countries (Greece, Ireland, and the UK) for that purpose (RUCAS, 2010). Infusing SD can also be achieved by involving environmental literacy and social responsibility in the curricula of multiple disciplines through an "across the curricula approach" that has been developed at a number of HEIs to help develop higher-order thinking skills such as critical thinking. The "across the curricula approach" starts by understanding environmental problems and potential solutions and developing social responsibility attitudes and skills required to help create a more humane and environmentally sustainable future (Rowe, 2002). Development of interdepartmental minors is another way as it provides for the opportunity to support cross-disciplinary work in environmental studies and to bring together students to share their diverse perspectives as they contribute to a common, integrated, closing course of this cluster. There is also a tendency to include CSR topics and courses in university and business school syllabi (Setó-Pamies et al., 2011; Wu et al., 2010) in addition to associated teaching methods (Hartman & Werhane, 2009; Seatter & Ceuleman, 2018). The main reason for such changes in courses and teaching pedagogies goes back to the impact these have on students' attitudes, beliefs, and values that change as a result of being involved in social responsibility learning (Luthar & Karri, 2005; Kleinrichert et al., 2011; Moon & Orlitzky, 2011).

Individual social responsibility (ISR) can be perceived as part of both CSR and USR. The main reason is that even if corporates or universities have their social responsibility strategies, without the actual individuals believing in their roles to take part in executing such strategies, little can be accomplished. ISR is defined as the responsibility of every individual for their actions (Park et al., 2009). It is the core essence of moral responsibility that directs how everyone should act. It defines the individual's level of commitment to the welfare of others; it functions best if individuals rise above self-interest (Ghemes, 2012, cited in Păceşilă, 2018). With such actions, each individual can start to make even small contributions to society and its natural resources (Corporate and Individual Social Responsibility, 2019).

Theoretical framework

To capture the complexity and contested concepts involved in this chapter, a combined set of different lenses is required. Therefore, the main theoretical framework that guides this topic builds on the Giving Voice to Values (GVV) (Gentile, 2010) and socio-cultural learning theory (Schunk, 2012). GVV is an innovative approach to values-driven leadership development that makes it easier for conversations about values and ethics to be integrated across the curriculum. GVV starts from the premise that most people already have values and want to act on them, but also want to feel that they have a reasonable chance of doing so effectively. This pedagogy focuses on building students' capability, and therefore confidence, to enact their values effectively. GVV is rooted in business and management disciplines and has "three As," where each A is reflected in a stage. The three As are "awareness," "analysis," and "action." The research presented in this chapter borrows from the GVV approach and its emphasis on the last step (namely action) by giving voice to students' values in the context of education for sustainable development as they ask themselves, "how do I most effectively give voice to my values?"

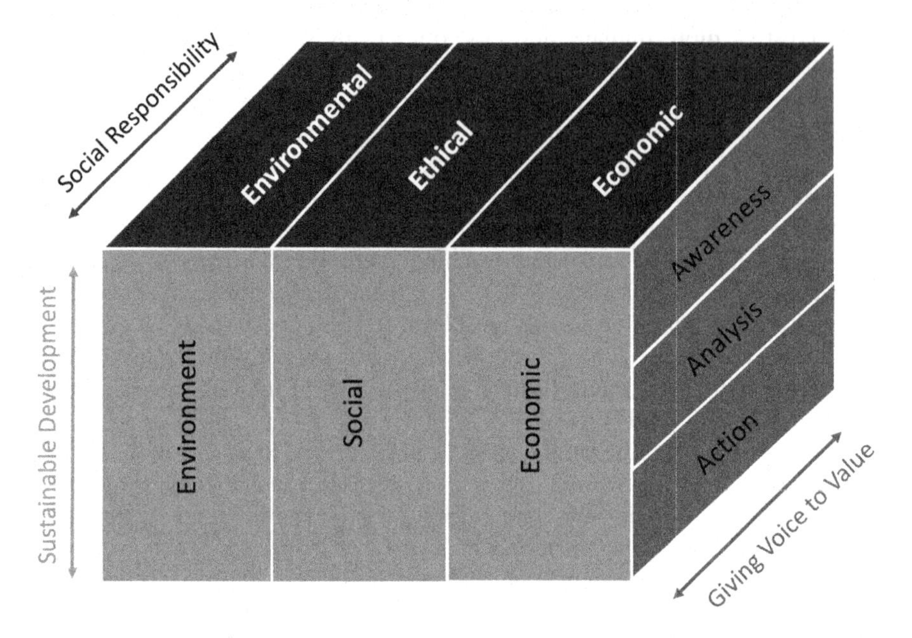

Figure 4.4 Theoretical framework.

GVV has been implemented in various contexts and cultures around the world. What makes it feasible is what Gentile (2018) highlighted as follows: GVV starts from respect and a set of shared values; it acknowledges context; it provides positive examples; and GVV is a metaphor for reframing choices and developing action plans. As for the socio-cultural theory, this is what guided the activities and the course design from the outset and throughout the various assignments, whether group or individual. The course was designed by utilizing a student-centered approach, where collaboration in group work and discussions were the base. The three main conceptual frameworks that guide this study are presented in Figure 4.4, where ESD, social responsibility, and GVV are illustrated in 3D to showcase how they intertwine within and among each dimension.

Research context

The study was conducted at the Graduate School of Education in a not-for-profit liberal arts private higher education institution in Egypt. The institution has a reputation of being one of the lead HEIs that caters for SD in its facilities and reduction of its carbon footprint, energy, water consumption, waste management, and efforts to raise campus awareness of environmental concerns (Office of Sustainability, 2019). This chapter illustrates the main milestones and processes that students went through during their enrolment in an elective

15-week course on ESD. Four cohorts were involved: three master of arts (MA) groups of students in 2015 ($n=10$), 2017 ($n=8$), and 2018 ($n=13$); and one undergraduate (UG) group of students in 2018 ($n=10$). Ethical approval of obtaining students' artefacts and assignments during the course were secured by institutional research board (IRB) approval. The course was designed based on the socio-cultural constructivist theory reflected in all course activities and assignments through student centred collaborative activities where knowledge is shared through social cultural tools and context (Johnson, 2009; Lave & Wenger, 1991). According to Derry (1999), in socio-cultural constructivism, culture and context in understanding what occurs in society and knowledge construction based on this understanding are emphasized. The course was designed in a modular format. Students were required to submit their "most significant changes" (MSC) after each of the five modules to describe what knowledge, skills, and values changed in them through reflective evidence that supported their claims. They also had to state what triggered this change and to identify clear futuristic plans and views on how this change will be taken further at the personal and professional levels. Figure 4.5 shows the five course modules and their topics.

The specific aim of this chapter is its focus on the final project, where students worked in groups through the course working hours and in many cases through additional working hours to design educational blogs. Their blogs required that they develop curricula material that included text, videos, and images in response to an authentic issue related to ESD in Egypt. Students' blogs were submitted as their final projects but were not tested in schools or with their target groups. For students to complete their blogs, they had to go through various stages, starting with deciding together after various iterations and discussions on a topic which they

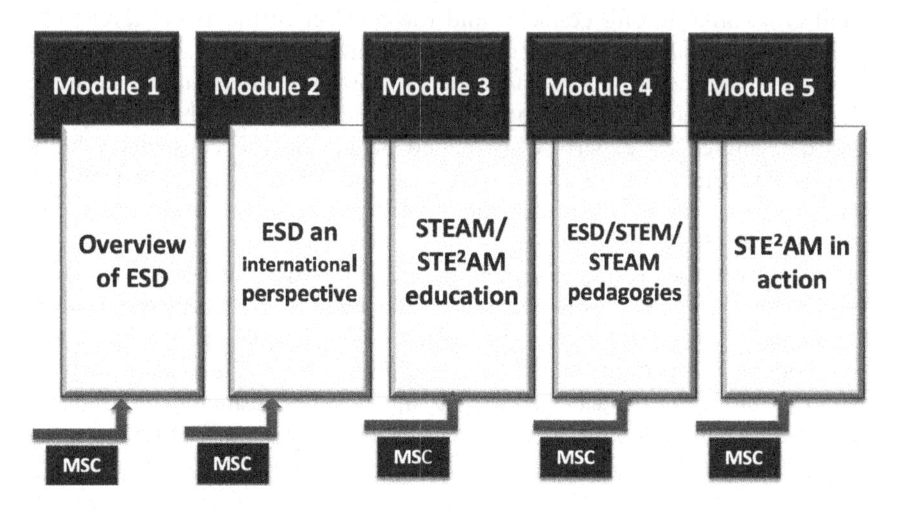

Figure 4.5 ESD course modules.

thought mirrored their areas of interests and values and reflected an SD contextual issue. The more general aim of this research and chapter is to explore how universities, as social institutions, can contribute to human and social development as a means to understanding universities' social responsibility.

Methods

Three main research questions guided this study: "What are students' perceptions of SD?"; "To what extent have students' understanding of ESD and its various dimensions developed throughout the course?"; "What values have students depicted in their final projects as a reflection of their social responsibility (SR) towards ESD?" To answer these questions, the study utilized a mixed-method approach. The study set out to investigate the possible impact of course involvement in enhancing students' understanding of ESD and how SR developed as an outcome. Three main instruments were designed to collect the data. These were the concept mapping rubric (CMR), most significant change (MSC), and end-of-course ESD project. To answer the first research question, participants were asked to construct a concept map to illustrate their understanding of the concept of sustainable development before engaging in the course content and activities, then they were asked to repeat that again at the end of the course. The aim was to probe on the cognitive organization students had before and after the course. Research by Novak and Gowin (1984) and Cronin et al. (1982) provides a general grading system for concept maps, but details related to ESD were not found. Therefore, a grading system was needed to analyze concepts under SD dimensions that are stated in the literature. A rubric reflective of a grading system was designed that analyzed the number of valid cross-links among concepts and the number of hierarchal levels. The reliability and validity of the rubric were determined by having a second rater. The two raters scored the concept maps and discrepancies between them were discussed until consensus was reached. As for the validity, this was determined by experts who helped decide on the construct validity. The lowest total score on the concept map was 3, while the highest was 12. The concept map tool was a quantitative instrument, where data was analyzed descriptively.

To answer the second research question, thematic analysis of students' responses to the MSCs were generated. The study utilized steps provided by Braun and Clark (2006), where the thematic analysis identified, analyzed, and reported on the patterns and themes within the data. These steps include: familiarizing with the data; generating initial codes; searching for themes, reviewing themes, defining themes, and finally writing-up. Each cohort was given a code (C1–C4) and each student was also coded depending on the total number of students per cohort (i.e., S1, S2, ...). The following present some of the quotations that were found during the analyses:

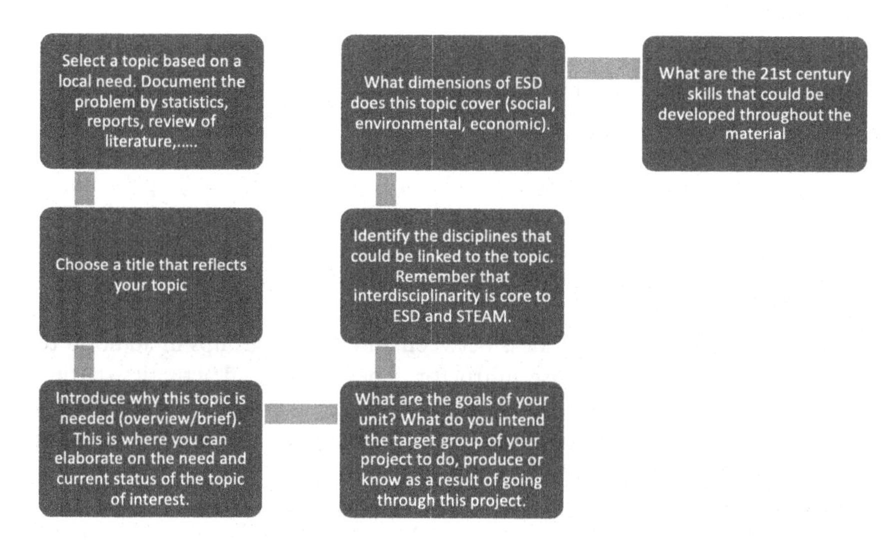

Figure 4.6 Stages of the ESD project development.

> C3S8: I have been through a transformational journey; it is mind-shifting. ESD taps into social, economic, as well as environmental dimensions. *ESD has grown on me to find its place in my head, hand, and heart.*

> C1S10: As a science teacher, it's my responsibility to incorporate issues of social justice, global citizenship, and ethics into my teaching. I have come to understand and appreciate how humanism can be incorporated into teaching the natural sciences, and look forward to doing so in the future. Most importantly, *I have found myself in the role of an "ESD ranger," without intentionally meaning to become one.*

> C2S5: My mind set has gone through a massive transition through this course. At the very beginning of the semester, I had no idea about sustainable development, now I see change in my mind-set and attitudes in a lot of my daily habits.

The third research question was answered by analyzing the final projects. Students followed the steps in Figure 4.6 to go about their projects. A qualitative rubric was designed to grade the projects that included the following criterion: topic of the project, originality and inventiveness, content that the project represents (ESD, 21st-century skills, science, technology, engineering, arts and maths (STEAM) disciplines), and media (videos and images) included in the project, and finally collaboration among the team members to contribute to share knowledge.

Findings and discussions

A total of 11 projects were developed, graded, and analyzed as students were perceived as empowered and responsible agents of change in their local societies. With such perception, students were actively engaged in learning about ESD as they perceived themselves as responsible citizens.

Each project was developed by three to four students working for three weeks at the end of the semester targeting an age range of their choice. The group members met frequently to discuss, decide, select, and synthesize material needed for their projects. Class time was devoted for students to continue their work and to present to the other groups on their progress. Students commented and shared ideas between and among groups in addition to receiving input from the course instructor. They developed activities starting with objectives and ending with assessment. The ESD course design provided a rich and open learning environment for students to voice their values and enact them while developing their projects. Although the course was not in the field of business, it enabled the students to work in a strictly free environment as they went through the stages of GVV. They had to be "aware" of the environmental, economic, cultural, and social issues in Egypt, and to "analyze" what are the contextual challenges related to the topic depicted in the project in order to "act" and develop an integrated curricular unit that could be implemented to a specific target audience as responsible, empowered

Table 4.1 WE garden

Middle stage school children, G6–9	
Type of education	*Extracurricular activity*
ESD dimensions	Economic: economics of gardening and recycling Social transformation: from school gardening to home gardening
Disciplines	Sciences (environmental biology) Social studies (mainly geography) Maths, ICT, arts & crafts, engineering, agricultural studies
Topics	Urban gardening, home gardening, soil composition, watering, selection and treatment of seedlings, timing of planting and harvesting, vertical gardens from home-made items Urban farming and creativity Composting helps students, recycling organic waste Carbon footprint, children's attention is drawn to the possible changes they can make to their lifestyle: diet, transportation, consumption and purchase of goods, recycling
Skills	21st-century skills Critical thinking, problem-solving, planning, communication and presentation skills, teamwork, ICT literacy, self-monitoring and peer monitoring, leadership
Values	Equity, fairness, diversity, change

Table 4.2 Character Education Camp (CEC): Learning skills for the future

NGO, youth, age 11–15 years old	
Type of education	*Non-formal*
Aim	Support youth to face social, economic, and environmental changes that influence global trends through an ESD character education
ESD dimensions	Environment, social, and economic
Disciplines	Technology/media, science, English language, engineering, arts
Topics	Social justice Humane being The environment
Skills	Critical and creative thinking, communication skills, awareness (self and others), reasoning, empathy, safety, problem-solving, self-reflection, teamwork, ICT literacy awareness, self-discipline, multicultural awareness, adaptability
Values	Equity, citizenship, social justice, tolerance

Table 4.3 Green culture elementary school

NGO, specializing in the provision of needs-based enrichment programs that infuse ESD into elementary school core curricula	
Aim	Reawaken the environmental consciousness Education for sustainable development (environmental, social) To incorporate interdisciplinary (STEEAM) within the existing curricula To nurture child development through 21st-century skills
Type of education	Informal education
ESD dimensions	Social and environmental
Disciplines	Physical education, science, English, geography, art
Topics	Organic farming, food security, child labour and food miles Sustainable agriculture Modern agriculture and the challenges it presents to sustainability Class-wide organic gardening project
Skills	Critical and creative thinking, the literacy domain develops media and technology literacy, and finally the psychosocial domain develops communication, initiative and collaboration skills Reflective thinking
Values	Child labour, right to education, responsibility

citizens making a change. The topics and target audience varied among students: child labor; traffic in Egypt; water and life; Character Education Camp (CEC): learning skills for the future; seasons in the sun; refugees and their rights; time machine; vocational education: building together a sustainable future; green culture; and social justice. Tables 4.1, 4.2, and 4.3 are examples of the analysis of three projects. In total, from the 11 projects, there were three projects directed for students in the primary stage, five projects for students in the elementary stage, and two for secondary stage students. There was one project for teachers and parents.

Discussion

The projects varied from those which could be start-up projects to those which could form possible partnerships with industry and decision-makers, whether corporates, businesses, NGOs, and/or ministries. The social responsibility students in both the MA and UG programs exhibited throughout their final projects how each project to increasing the welfare of humanity (Amoako et al., 2015). Values evolving from the final projects centered around those that involve humane aspects, such as human dignity, human rights, equity, respect for diversity, and economic justice. The research perceives that it builds on the Giving Voice to Values approach (Gentile, 2010). This is mainly as students started their courses with tacit orientations that drove their project choices and discussions inside and outside the classes. The course provided students the chance to enact the ESD 2 type (Vare & Scott, 2007), as the focus was on the learning process and student capacity-building through reflection, decision-making, and critical thinking while in their groups. The course also seemed to have impacted their academic and social domains as a result of the experiences they went through (Conrad & Hedin, 1982; Scales et al., 2000). As a result of this course, as part of the USR, there could be an argument that achieving Vision 2030 and the SDGs could be possible through educational efforts such as those by HEIs. The findings of this study align with other studies related to USR, such as Alzyoun and Bani-Hani (2015), Amoako et al. (2015), Brondani (2012), Filho et al. (2019), Grau et al. (2017), which were carried out in the region and elsewhere.

Conclusion

The chapter presented students' experience while enrolled in an elective ESD course at the graduate and undergraduate levels. During their enrollment, students submitted group projects that could be seen as an opportunity for implementing a GVV approach through developing social responsibility among students. This study claims that it developed students' understanding of social responsibility as it evolved throughout the course. This understanding seemed to evolve as they synthesized their understanding of ESD. In conclusion, the course design provided an opportunity for students to voice their values and develop actionable projects to overcome contextual issues related to ESD in Egypt. The study presented here fills a gap related to SD in Egypt by providing an authentic experience of HE students related to ESD practices in the context of a course in education. The projects students developed can help develop the economic and commercial impact of HE, whereas some of the projects opened new job opportunities or entrepreneurial projects. HEIs could therefore be a place with a concern about social transformation as "universities cannot be sustainable without being socially responsible" (Mahoney, cited in Weiss, 2016).

Implications in business education

The findings from this research can inform business education by different means, as identified in the literature (i.e., Casser, 2019; Painter et al., 2018; Tormo-Carbó et al., 2018; Wright & Bennett, 2011). Courses in business education can include practical implications in the delivery of integrated cases that combine intended coursework and ethics. Having students enrolled in business education in higher education institutions in an active participant role as courses such as those with an emphasis on sustainable development can influence ethical awareness where the development and implementation of sustainability practices may be facilitated. Infusing sustainable ethics in business courses can be useful and a potential critical component to any curriculum that intends to prepare future professionals to be effective contributors to a sustainable society. Such impact can expand in the personal and professional lives of participating students while also impacting businesses and corporations with employees with a solid ethical foundation for making more sustainable-oriented decisions. As for the nature of courses related to sustainable development in business education, they can be offered as elective courses, where each course integrates main components linked to ethical, economic, and environmental issues. MBA programs could also include aspects related to CSR, sustainability, and business ethics as they are seen to be of great interest in management education from the perspectives of potential employers, alumni, and various accrediting bodies in higher education.

Recommendations

Based on the findings from this study, several recommendations can be made to showcase the impact of university and individual social responsibility as they are developed through intended coursework by university faculty and students. Viewing this study from both an economic and commercial impact lens sets the following recommendations: integrate social responsibility into university and schools mission statements to become an integral part of what they do; encourage more community engagements and outreach through universities partnering with local and international organizations that help maximize their impact on society; clarify the university's identity with one that speaks to the social and public mission of the country's national context; support and equip students in higher education institutions to become solvers of society's problems. By promoting USR, not only does it demonstrate the university's commitment to social responsibility practices, but it helps gain a public image as it improves the university's reputation, its ranking, and helps attract further funding, impacts key stakeholders, and supports better research productivity. As a result, graduates of HEIs become good corporate citizens (Nagy & Robb, 2008) with an impact on their societies as they strengthen their civil commitment and active citizenship as advocates for social change.

References

Alikihc, O. (2015). Broadening the concept of green marketing: Strategic corporate social responsibility. In A. Ulas, *Handbook of Research on Developing Sustainable Value in Economics, Finance and Marketing*. Hershey, PA: IGI Global.

Alzyoun, S., & Bani-Hani, K. (2015). Social responsibility in higher education institutions: Application case from the Middle East. *European Scientific Journal, 11*(8), 122–129.

Amin-Chaudhry, A. (2016). Corporate social responsibility – from a mere concept to an expected business practice. *Social Responsibility Journal, 12*(1), 190–207. https://doi.org/10.1108/SRJ-02-2015-0033.

Amoako, G. Agbola, R. Dzogbenuku, R, & Sokro, E. (2015). CSR and education: The Ghanaian and African perspective. *Education and Corporate Social Responsibility International Perspectives*, 185–222. https://doi.org/10.1108/S2043-0523(2013)0000004011.

Aupperle, K. E. (1984). An empirical measure of corporate social orientation. In L. E. Preston (ed.). *Research in Corporate Social Performance and Policy*, 6 (JAI, Greenwich, CT), pp. 27–54.

Bozkurt, F., & Ergen, A. (2015). Promoting healthy lifestyle for sustainable development. In A. Ulas, *Handbook of Research on Developing Sustainable Value in Economics, Finance and Marketing*. Hershey, PA: IGI Global.

Braun, V., & Clarke, V. (2006). Using thematic analysis in psychology. *Qualitative Research in Psychology, 3*, 77–101.

Brondani, M. A. (2012). Teaching social responsibility through community service-learning in predoctoral dental education. *Journal of Dental Education, 76*(5), 609–619.

Brunelle, J. P. (2001). *The Impact of Community Service on Adolescent Volunteers' Empathy, Social Responsibility, and Concern for Others*. Doctoral dissertation, Virginia Commonwealth University.

Camilleri, M. A. (2017a) *Corporate Sustainability, Social Responsibility and Environmental Management: An introduction to theory and practice with case studies*. Heidelberg, Germany: Springer.

Camilleri, M. A. (2017b) Corporate sustainability and responsibility: Creating value for business, society and the environment. *Asian Journal of Sustainability and Social Responsibility, 2*(16), 59–74. https://doi.org/10.1186/s41180-017-0016-5.

Casser, C. (2019). Business ethics and sustainable development. In W. Leal Filho (ed.) *Encyclopedia of Sustainability in Higher Education*. Cham: Springer. Retrieved from https://link.springer.com/referenceworkentry/10.1007%2F978-3-319-63951-2_39-1.

Conrad, D., & Hedin, D. (1982). Corporate sustainability and responsibility: Creating value for business, society and the environment: The impact of experimental education on adolescent development. *Child & Youth Services, 4*(3–4), 57–76.

Corporate and Individual Social Responsibility (2019). 123HelpMe.com. Retrieved from: www.123helpme.com/view.asp?id=243848.

Cronin, P., Dekker, J., &, Dunn, J. (1982). A procedure for using and evaluating concept maps. *Research in Science Education, 12* (1), 17–24.

Derry, S. J. (1999). A fish called peer learning: Searching for common themes. *Cognitive Perspectives on Peer Learning*, 197–211.

DeShields, O. W., Jr., Kara, A., & Kaynak, E. (2005). Determinants of business student satisfaction and retention in higher education: Applying Herzberg's two-factor theory. *International Journal of Educational Management, 19*(2), 128–139.

EL-Deghaidy, H. (2012). Education for sustainable development: Experiences from action with science teachers. *Discourse and Communication for Sustainable Education, 3*, 23–40.

Elkington, J. (1997). *Cannibals with Forks: The triple bottom line of 21st century business.* Oxford: Capstone.

Filho, N. (2017). *University Social Responsibility Indicators System Sharing Information on Progress – PRME: Guide for implementation.* UN Global Compact, PRME Regional Chapter Latin America and the Caribbean. Retrieved from www.unprme.org/resource-docs/cartillaprimeinglsonline.pdf.

Filho, W., Doni, F., Vargas, V., Wall, T., Hindley, A., Rauman-Bacchus, L., Emblen-Perry, K., et al. (2019). The integration of social responsibility and sustainability in practice: Exploring attitudes and practices in higher education institutions. *Journal of Cleaner Production, 220*, 152–166.

Furco, A. (1994). A conceptual framework for the institutionalization of youth service programs in primary and secondary education. *Journal of Adolescence, 17*(4), 395.

Gallardo-Vazquez, D., & Sanchez-Hernandez, M. (2014). Measuring corporate social responsibility for competitive success at a regional level. *Journal of Cleaner Production, 72*, 14–22. doi: 10.1016/j.jclepro.2014.02.051.

Gentile, M. (2010). *Giving Voice to Values: How to speak your mind when you know what's right.* New Haven, CT: Yale University Press. Curriculum site: www.GivingVoiceToValues.org.

Gentile, M. (Sept. 2018). Giving voice to values. Stanford Social Innovation Review: Informing and inspiring leaders of social change. https://ssir.org/articles/entry/giving_voice_to_values.

Grau, F., Escrigas, C., Goddard, J., Hall, B., Hazelkorn, E., & Tandon, R. (2017). Towards a socially responsible higher education institution: Balancing the global with the local. In F. Grau, B. Hall, & R. Tandon (eds.) *Higher Education in the World 6: Towards a socially responsible university: Balancing the global with the local.* GUNi Series on the Social Commitment of Universities, 37–51.

GRI (nd). *GRI Standards.* Global Strategic Alliances. www.globalreporting.org/Pages/default.aspx.

Hartman, L., & Werhane, P. H. (2009). A modular approach to business ethics integration: At the intersection of the stand-alone and the integrated, *Journal of Business Ethics, 90*, 295–300.

Hopkins, M. (2005). Measurement of corporate social responsibility. *International Journal Management and Decision Making, 6*(3/4), 213–231.

Johnson, K. E. (2009). *Second Language Teacher Education: A sociocultural perspective.* New York: Routledge.

Kleinrichert, D., Albert, M., & Eng, J. P. (2011), The role of corporate values on business students' attitudes: A comparison of undergraduates and MBAs. *The Business Review, 17*(1), 53–59.

Kumar, S., Kumar, N., & Vivekadhish, S. (2016). Millennium Development Goals (MDGs) to Sustainable Development Goals (SDGs): Addressing unfinished agenda and strengthening sustainable development and partnership. *Indian Journal of Community Medicine, 41*(1), 1–4. doi: 10.4103/0970-0218.170955.

Landorf, H., Doscher, S., & Rocco, T. (2008). Education for sustainable human development: Towards a definition. *Theory and Research in Education, 6*(2), 221–236.

Lave, J., & E. Wenger. (1991). *Situated Learning: Legitimate peripheral participation.* Cambridge: Cambridge University Press.

Luthar, H., & Karri, R. (2005). Exposure to ethics education and the perception of linkage between organizational ethical behavior and business outcomes. *Journal of Business Ethics, 61*(4), 353–368.

Maignan, I., & Ferrell, O. C. (2000). Measuring corporate citizenship in two countries: The case of the United States and France. *Journal of Business Ethics, 23*(3), 283–297.

Malone, D. (2018). United Nations University's introduction. In F. Grau, B. Hall, & R. Tandon (eds.), *Higher Education in the World 6: Towards a socially responsible university: Balancing the global with the local*, GUNi Series on the Social Commitment of Universities, 33–34.

McKeown, R., & Hopkins, C. (2003). EE≠ ESD: Diffusing the worry. *Environmental Education Research, 9*, 117–128.

McMillin, J., & Dyball, R. (2009). Developing a whole-of-university approach to educating for sustainability linking curriculum, research and sustainable campus operations. *Journal of Education for Sustainable Development, 3*(1), 55–64. https://journals.sagepub.com/doi/pdf/10.1177/097340820900300113.

MDG (n.d.). *Millennium Development Goals.* Retrieved from www.mdgfund.org/content/MDGs.

Ministry of International Cooperation (2016). *National Voluntary Review on the Sustainable Development Goals: Input to the 2016 High-level Political Forum (HLPF) on Sustainable Development.* The Arab Republic of Egypt. Retrieved from https://sustainabledevelopment.un.org/content/documents/10738egypt.pdf.

Ministry of Planning, Monitoring and Administrative Reform (2018). *Egypt's Voluntary National Review 2018.* Arab Republic of Egypt, 2030 Vision of Egypt.

Moon, J., & Orlitzky, M. (2011). Corporate social responsibility and sustainability education: A Trans-Atlantic comparison. *Journal of Management and Organization, 17*(5), 583–603.

Nagy, J., & Robb, A. (2008). Can universities be good corporate citizens? *Critical Perspectives on Accounting, 19*(8), 1414–1430.

Newmann, F. M., & Rutter, R. A. (1983). *The Effects of High School Community Service Programs on Students' Social Development.* Madison, WI: Wisconsin Center for Education Research.

Novak, J., & Gowin, B. (1984). *Learning How to Learn.* Cambridge: Cambridge University Press.

Office of Sustainability (2019). *Sustainable Campus.* www.aucegypt.edu/about/sustainable-auc/sustainable-campus.

Othman, S. & EL-Deghaidy, H. (2008). Relationship between aesthetic values and environmental peace: An absent aspect in the national standards of education in the curricula: a suggested proposal. *Journal of Educational Sciences, 1*, 295–333.

Păceşilă, M. (2018). The individual social responsibility: Insights from a literature review. *Management Research and Practice, 10*(1), 17–26. Retrieved from http://mrp.ase.ro/no101/f2.pdf.

Painter, M., Hinnert, S., & Cooper, T. (2018). The development of responsible and sustainable business practice: Value, mind-sets, business-models. *Journal of Business Ethics*, *157*(4), 885–891.

Park, H. S., Shin, Y. S., & Yun, D. (2009). Differences between White Americans and Asian Americans for social responsibility, individual right and intentions regarding organ donation. *Journal of Health Psychology*, *14* (5), 707–712.

Quazi, A. M., & O'Brien, D. (2000). An empirical test of a cross-national model of corporate social responsibility. *Journal of Business Ethics*, *25*, 33–51.

Raskoff, S. A., & Sundeen, R. A. (1999). Community service programs in high schools. *Law and Contemporary Problems*, *62*(4), 73–111.

Rowe, D. (2002). Environmental literacy and sustainability as core requirements: Success stories and models. In W. Leal Filho (ed.), *Teaching Sustainability at Universities: Towards curriculum greening*. Frankfurt: Peter Lang, 79–103.

RUCAS (2010). http://rucas.edc.uoc.gr/rucas.

Sammalisto, K., & Arvidsson, K. (2005). Environmental management in Swedish higher education: Directives, driving forces, hindrances, environmental aspects and environmental co-ordinators in Swedish universities. *International Journal of Sustainable Higher Education*, *6*(1), 18–35.

Salem, H., El-Maghrabi, M. H., Rodarte, I., & Verbeek, J. (2018). *Sustainable Development Goal Diagnostics: The case of the Arab Republic of Egypt*. Policy Research Working Paper 8463. World Bank Group: Office of the Senior Vice President UN Relations and Partnerships. Retrieved from http://documents.worldbank.org/curated/en/532831528165791465/pdf/WPS8463.pdf.

Sanchez-Hernandez, M., & Mainardes, E. (2016). University social responsibility: A student base analysis in Brazil. *International Review Public Nonprofit Mark*, *13*, 151–169. doi: 10.1007/s12208-016-0158-7.

Scales, P. C., Blyth, D. A., Berkas, T. H., & Kielsmeier, J. C. (2000). The effects of service learning on middle school students' social responsibility and academic success. *Journal of Early Adolescence*, *20*(1), 332–358.

Schunk, D. (2012). *Learning Theories: An educational perspective* (6th ed.). Boston: Pearson Education, Inc.

Seatter, C., & Ceuleman, K. (2018). Teaching sustainability in higher education: Pedagogical styles that make a difference. *Canadian Journal of Higher Education*, *47*(2), 47–70.

Setó-Pamies, D., Domingo-Vernis, M., & Rabassa-Figueras, N. (2011). Corporate social responsibility in management education: Current status in Spanish universities. *Journal of Management & Organization*, *17*(5), 604–620.

Setó-Pamies, D., & Papaoikonomou, E. (2016). A multi-level perspective for the integration of ethics, corporate social responsibility and sustainability in management education. *Journal of Business Ethics*, *136*, 523–538.

Smith D. (1993). The Frankenstein syndrome: Corporate responsibility and the environment. In D. Smith (ed.), *Business and the Environment: Implications of the new environmentalism*. London: Paul Chapman, pp. 172–189.

Temkar, K. (n.d.). *Corporate Social Responsibility Activities of Top Selected Listed Companies*. A research proposal submitted to the Swami Ramanand Teerth Marathwada University, Nanded. Retrieved from https://shodhgangotri.inflibnet.ac.in/bitstream/123456789/7073/1/reseach%20propsal%20kiran%20temkar.pdf.

Thomas, G., & Nowak, M. (2006). *Corporate Social Responsibility: A definition*. Graduate School of Business, Curtin University of Technology.

Tilbury, D., & Wortman, D. (2004). *Engaging People on Sustainability*. Gland, Switzerland and Cambridge, UK: Commission on Education and Communication, IUCN.

Tormo-Carbó, G., Seguí-Mas, E., & Oltra, V. (2018). Business ethics as a sustainability challenge: Higher education implications. *Sustainability*, *10*, 2717. Retrieved from https://doi.org/10.3390/su10082717.

Turker, D. (2008). Measuring corporate social responsibility: A scale development study. *Journal of Business Ethics*, *85*, 411–427. doi: 10.1007/s10551-008-9780-6.

UNESCO (2005). *Education for Sustainable Development: An expert review of processes and learning*. Retrieved from http://unesdoc.unesco.org/images/0019/001914/191442e.pdf.

UNESCO (2008). *Regional Guiding Framework of Education for Sustainable Development in the Arab Region*. Education for Sustainable Development. United Nations Decade (2005–2014). Beirut: UNESCO Regional Bureau for Education in Arab States. Retrieved 3 August 2014 from www.esd-world-conference-2009.org/fileadmin/download/general/Arab_ESD_regional_strategie.pdf.

UNESCO (2014a). *Shaping the Future We Want: UN Decade of Education for Sustainable Development (2005–2014)*. Final Report. http://unesdoc.unesco.org/images/0023/002301/230171e.pdf.

UNESCO (2014b). *UNESCO Roadmap for Implementing the Global Action Programme on Education for Sustainable Development*. Paris: UNESCO.

United Nations [UN] (n.d.). *Key Messages of Egypt VNR 2018*. Retrieved from https://sustainabledevelopment.un.org/memberstates/egypt.

United Nations [UN] (2005). *UN Decade of Education for Sustainable Development 2005–2014*. https://unesdoc.unesco.org/ark:/48223/pf0000141629.

United Nations [UN] (2011). *The Millennium Development Goals Report 2011*. Sales No. E.11.I.10.

Vare, P., & Scott, W. (2007). Learning for a change: Exploring the relationship between education and sustainable development. *Journal of Education for Sustainable Development*, *1*(2), 191–198.

Wade, R. C. (1997). *Community Service Learning: A guide to including service in the public school curriculum*. New York: SUNY Press.

Wals, A., & Blaze Corcoran, P. (2006). Sustainability as an outcome of transformative learning. Education for Sustainable Development in Action, Technical Paper No. 3. In J. Holmberg, & B. Samuelsson (eds.), *Drivers and Barriers for Implementing Sustainable Development in Higher Education*. Paris: UNESCO.

Wals, A., & Kieft, G. (2010). *Education for Sustainable Development: Research overview*. Stockholm: SIDA.

Weiss, B. (August, 2016). The rise of social responsibility in higher education. *University World News: The Global Window on Higher Education*. Retrieved from www.universityworldnews.com/post.php?story=20160811095808959.

Wells, P. (2017). UNESCO'S Introduction: The role of higher education institutions today. In F. Grau, B. Hall, & R. Tandon (eds.), *Higher Education in the World 6: Towards a socially responsible university: Balancing the global with the local*. GUNi Series on the Social Commitment of Universities, 31–32.

Windsor, D. (2001). The future of corporate social responsibility *International Journal of Organizational Analysis*, *9*(3), 225–256. doi: 10.1108/eb028934.

Wood, D. J., & Jones, R. E. (1995). Stakeholder mismatching: A theoretical problem in empirical research on corporate social performance. *International Journal of Organizational Analysis*, *3*, 229–267.

World Commission on Environment and Development [WCED] (1987). *Our Common Future*. Oxford: Oxford University Press.

World Economic and Social Survey (2013). *Sustainable Development Challenges*. Department of Economic and Social Affairs. New York: United Nations. E/2013/50/Rev.1 ST/ESA/344. Retrieved from https://sustainabledevelopment.un.org/content/documents/2843WESS2013.pdf.

Wright, N., & Bennett, H. (2011). Business ethics, CSR, sustainability and the MBA. *Journal of Management and Organization*, *17*(5),641–655.

Wu, Y., Huang, S., Kuo, L., & Wu, W. (2010). Management education for sustainability: A web-based content analysis. *Academy of Management Learning and Education*, *9*(3), 520–531.

5 Business ethics in the business schools in Morocco

Wafa El-Garah and Asmae El Mahdi

Introduction

Corporate ethics violations continue making the headlines. The proliferation of scandals, in recent years, seems to have increased, with new scandals revealed every year. Table 5.1 lists the top corporate scandals exposed in 2018. When discussing the underlying cause of these scandals, often the fingers are pointed to business schools over their approach, in teaching future business managers and leaders, overemphasizing the bottom line at all costs (Melé 2008). Business schools have long been criticized for promoting profit maximization and shareholder value over stakeholders' interest (Gioia 2002; Ronald & Ragatz 2008). As stated by Freeman et al. (2009), "Teach business students that only shareholders matter – taking precedence over employees, communities, and suppliers – and they will go forth to give us the current financial crisis" (p. 39).

In response to these accusations, business schools have rushed into crafting different solutions to address the issue. They have revised their missions to incorporate corporate social responsibility and ethical behavior (Davis et al. 2007; Holosko et al. 2015, Lopez & Martin 2018). They have also updated their curricula to include concepts of ethics and corporate social responsibility (Buff & Yonkers 2004). In fact, after the Enron scandal, Tran (2015) found that the number of business ethics courses in business curricula has increased by 61%. Business schools have adopted different approaches (Park 1998; Sims & Felton 2006; Cornelius et al. 2007); some have added a new course on business ethics, sustainability, corporate social responsibility, and environmental issues; others have taken an integrative approach by embedding these concepts into different courses spanning the different levels of the programs (Schoenfeldt et al. 1991; Roberts & Roper 2015; Sauser & Sims 2015). Concomitantly, new innovative approaches to teaching ethics have emerged, such as the Giving Voice to Values (GVV) curriculum (Gentile 2010; www.GivingVoiceToValues.org). Additionally, the United Nations launched global movements such as the UN Global Compact and the Principles for Responsible Management Education (PRME) (Tavanti & Wilp 2015). Both the Global Compact and PRME are voluntary pacts to encourage businesses and business schools worldwide to adopt principles of sustainability and social responsibility, among others.

Table 5.1 The top corporate scandals in 2018

Organization	*Scandal*
Facebook and Cambridge Analytica	Over 80 million Facebook users' data was used by a third party without their consent. Allegedly the data was used unethically by Cambridge Analytica, a company hired by Donald Trump during his election campaign. Using a controversial technique called psychographic modeling, the company could exploit the private data of Facebook users to affect voters' behavior (Wolff-Mann, 2018).
Renault–Nissan–Mitsubishi Alliance	Carlos Ghosn, chairman and CEO of Renault–Nissan–Mitsubishi Alliance, was arrested over financial misconduct. He allegedly misused company assets and underreported his income between 2010 and 2015 (Wolff-Mann, 2018).
Deutsche Bank	On November 29, 2018, Deutsche Bank offices in Frankfurt were raided by police over money laundering allegations. The bank seems to have had a role in a US$20 billion Russian money laundering scheme (Harding, 2019).
Goldman Sachs	Goldman Sachs bankers have been charged with bribery and money laundering due to their involvement in the 1Malaysian Development Berhad (1MBD) fraud (1MBD is a Malaysian state-owned development fund (Wolff-Mann, 2018).
Tesla	Tesla and its CEO Elon Musk were each given a US$20 million fine by the Security Exchange Commission (SEC) following a tweet by Elon Musk that he had secured funding to take the company private at $420 per share, which was later found to be a lie (Wolff-Mann, 2018).
Google	Google executives decided to keep a data breach incident of Google+ from the public. Due to a software malfunction, an estimated 500,000 Google+ users' information was compromised. Shortly after this information was revealed, Google announced the shutting down of the Google+ platform (Wolff-Mann, 2018).

Accreditation bodies, the advocates of quality in management education, have led the movement and are aligning their standards with the urgent need to tackle ethical and corporate social responsibility issues. In fact, accrediting agencies such as the European Foundation for Management Development (EFMD) and the Association to Advance Collegiate Schools of Business (AACSB), which are regarded as the most prominent business school accrediting agencies worldwide, have revised their standards, putting more emphasis on corporate social responsibility, sustainability, and ethics (EFMD 2019; AASCB 2018). Additionally, EFMD developed a new tool, Business School Impact System (BSIS), that helps schools and universities assess their direct and indirect impact locally and globally (EFMD 2019).

Research addressing the issues of business ethics in education has also increased. A simple search of "business ethics" in the EBSCO host database revealed 104,575 results, corresponding to 99,515 academic journals, 3,090 reviews, 651 conference materials, 138 reports, and 124 books. Whereas a decade ago, Schwartz (2008, p. 218) found only 11,000 hits in the ABI/Inform database. Nevertheless, most of these research outputs rest on data collected from developed countries. Studies covering developing countries remain scarce. Despite the fact that most scandals making headlines emanate from developed countries such as the US (Enron, WorldCom, Facebook), Germany (Volkswagen, Siemens), France (Airbus, Renault), the UK (Cambridge Analytica, Rolls Royce), to name only a few, developing countries are plagued with challenging business ethics issues, namely corruption and fraud, which seldom make it to the international news headlines. Some research studies (Simpson et al. 2015; Onumah et al. 2012) have addressed business ethics education in accounting programs in Ghana, a developing country in Africa. Simpson et al. (2015) reported that private, religious educational institutions tend to offer business ethics courses more than their public, non-religious counterparts.

Hence, the purpose of this chapter is to attempt to fill this gap by surveying the current situation of business ethics education in higher education institutions in Morocco, particularly within business schools offering business and management programs. This chapter is organized as follows: first, a brief overview of the evolution of business ethics education and the role of PRME are presented. Second, the ethical landscape in Morocco's business environment is described. Third, the new approach to teaching business ethics, Giving Voice to Values (GVV), is outlined. Fourth, the results of an exploratory study within Moroccan business schools are discussed. Fifth, the descriptive case study of Al Akhawayn University in Ifrane, a liberal arts public Moroccan higher education institution, is presented to highlight how it is leading the way in mainstreaming business ethics education within its school of business administration.

Evolution of business ethics as a field

Business ethics has undergone a remarkable development journey. Many have attempted to trace the rise and development of business ethics (DeGeorge 1987; Abend 2013, 2014, 2016; Tran 2015; Bird 2016). DeGeorge is regarded as the precursor in differentiating business ethics from other streams of study (Schwartz 2008). DeGeorge (1987) reveals five stages leading to the gradual development and consolidation of business ethics. Before the 1960s, business ethics was in the phase of "ethics in business." Accordingly, "ethics in business is not new. It is as old as business" (DeGeorge 1989). Origins are found to extend as far back as 4,000 years ago, when the Hammurabi Code was established in Mesopotamia in an attempt to arrange for "honest price" (Schwartz 2008, p. 219). Intellectual roots go as far back as Aristotle, who discussed the moral attributes of merchants. The Jewish Talmud and Biblical doctrine provided guidelines on proper business conduct, while the Koran emphasized business ethics regarding matters of poverty and wealth (Schwartz 2008, p. 216). Ideas further bloomed with the work of thinkers in the 17th and 18th centuries, namely Immanuel Kant, John Stuart Mill, and Adam Smith, who discussed the link between ethics and business (Tran 2015). During this phase, business practice and ethics were widely considered as being unrelated, for business was regarded as "an amoral activity" (Schwartz 2008, p. 217).

Starting from the 1960s, business ethics started taking shape. In response to the rise of the 1960s social issues, including civil rights, ecological issues arising from industries' development, workers' rights, and consumers' concerns, business schools started introducing "social issues courses" (DeGeorge 1987; Tran 2015). Management professors designed the courses with material initially skewed in favor of managers' perspectives; this was subsequently adjusted by incorporating alternative perspectives of consumers and workers. Business schools' responses restricted the courses' scope on the social responsibilities of business within the boundaries of law and "what is legal" (DeGeorge 1987). Professors from different disciplinary backgrounds convened in meetings and conferences to discuss business practice and its emerging responsibilities (DeGeorge 1987).

The 1970s saw the entry of philosophers to the area of business ethics. Such an entry was previously challenged: "Some professors of management questioned the role – if any – philosophers could play in an area with which philosophers were unacquainted and towards which many of them were hostile" (DeGeorge 1987). Changing circumstances paved the way for philosophers' entry, including the rise of biomedical ethics coupled with the wide outcry regarding the series of scandals and misconduct revealed in media coverage, which raised public awareness and aroused intellectual interest among students (DeGeorge 1987). In addition to these changing realities, DeGeorge (1987) argues that "Rawls's *A Theory of Justice* legitimized philosophical concern with economic issues." Accordingly, business ethics started emerging as a "field" generating work centered around new issues revolving around the "moral status of the Corporation" (DeGeorge 1987).

The 1980s saw business ethics emerging as a recognizable interdisciplinary academic field thanks to the institutionalization groundwork (DeGeorge 1987). In addition to emerging knowledge generation, in different formats, including textbooks, casebooks, publications, and conferences featuring over 40 scholars, 500 courses taught business ethics to 40,000 students across business schools in the US, including MBA programs in Harvard, coupled with emerging attempts of training business professors on theories of ethics, namely the University of Kansas's 12 professors (DeGeorge 1987). The institutionalization movement faced questions of whether business ethics can be taught (Henderson 1988) and transcended the academic domain to reach corporate life; business ethics started emerging in business practice with the rise of "ethics committees" and "social policy committees" within firms

(DeGeorge 1987). DeGeorge (1987) explains the significance of defining the field of business ethics:

> Defining business ethics as a field helps distinguish it from the polemical writing in the area. It also helps emphasize the interrelatedness of the problems with which it deals and of the several levels of discourse which it embraces. Distinguishing it as a field does not deny its partial but significant overlap with its earlier antecedents – ethics in business and corporate social responsibility; nor does it deny its connections with general ethics, other areas of applied ethics, and the various fields of business education.

1985 marked the fifth development stage of the business ethics field. Schwartz (2008, p217) categorizes this stage into three major developments. The 1985–1995 period saw growing organizational changes within corporations by developing codes of conduct, integrating ethics officers, and adopting training in ethics and ethics hotlines. The 1995–2000 period featured the rise of issues from international business, notably bribery and corruption of public servants and suppliers' use of child labor. The current period, starting from 2000, has witnessed corporate scandals, notably Enron, WorldCom, and Tyco (Schwartz 2008; Mason 2018, p. 1858; Holland & Albrecht 2013).

In response to corporate scandals, the 2000s witnessed the introduction of global initiatives and legislative reforms to enforce business ethics. In 2000, the United Nations established the UN Global Compact, which is the world's largest corporate sustainability initiative (unglobalcompact.org). It is a voluntary pact that corporate CEOs commit to aligning their corporate strategies and operations with 10 sustainability principles. These 10 principles are listed in Table 5.2. Following the many corporate accounting scandals in major US organizations, such as Enron and WorldCom, the US Congress passed the Sarbanes–Oxley Act in 2002. This legislation aims to enhance the integrity of financial statements and audits. The Act also mandates several policies on US publicly traded companies with regards to corporate governance practices. According to Kollar and Green (2008, p. 1851), the Sarbanes–Oxley Act is widely regarded as the most significant legislation since the Securities and Exchanges Act of 1933 and 1934. While debates have raged over the cost benefit of Sarbanes–Oxley (Ribstein 2002; Leuz 2007; Ugrin & Odom 2010), other countries, namely France, Japan, Canada, and China, have passed their own versions of the Sarbanes–Oxley Act in an attempt to curb corporate scandals (Zhang & Han 2016).

Corporate scandals, along with recurrent economic crises, have had other sweeping implications. The wave of scandals leading to scrutiny with the conviction and imprisonment of many corporate managers (Khurana 2007) led to "undermining confidence in capital markets, eroding trust in professional institutions, and casting a shadow over the probity of corporate life and those institutions affiliated with it – including business schools"

Table 5.2 The 10 principles of the UN Global Compact

Human rights	Principle 1: Businesses should support and respect the protection of internationally proclaimed human rights; and Principle 2: make sure that they are not complicit in human rights abuses.
Labor	Principle 3: Businesses should uphold the freedom of association and the effective recognition of the right to collective bargaining; Principle 4: the elimination of all forms of forced and compulsory labor; Principle 5: the effective abolition of child labor; and Principle 6: the elimination of discrimination in respect of employment and occupation.
Environment	Principle 7: Businesses should support a precautionary approach to environmental challenges; Principle 8: undertake initiatives to promote greater environmental responsibility; and Principle 9: encourage the development and diffusion of environmentally friendly technologies.
Anti-corruption	Principle 10: Businesses should work against corruption in all its forms, including extortion and bribery.

Source: www.unglobalcompact.org/what-is-gc/mission/principles.

(Adler 2002). Accordingly, the scandals catalyzed a wave of criticism on business schools, "reopening the question of what exactly this institution is for, what functions we as a society want it to perform, and how well it is performing them" (Khurana 2007, p. 5). Criticized for producing managers solely focused on shareholder value in their pursuit of short-term goals for profit maximization at the expense of stakeholders' values, business schools' approach proved to be a "recipe for disaster" (Currie et al. 2010). Business schools saw mounting pressure to address their educational shortcomings and adapt to changing realities towards producing managers equipped to better handle the human and social complexities of business practice (Bennis & O'Toole 2005).

The scandals' worrisome magnitude brought the need for business ethics into the limelight, leading to its growing presence. Companies such as Citigroup, who found themselves implicated in legal breaches in the wake of corporate scandals, required its employees to take mandatory, online ethics courses (Tran 2015). Hollywood made movies depicting and tackling issues in business ethics, namely "*Wall Street, Quiz Show, Boiler Room, Erin Brockovich, The Insider,* and *Jerry Maguire* and even in children's films such as *Monsters, Inc*" (Schwartz 2008, p. 217). Additionally and after the Enron scandal, a survey of 91 business schools in the US found an increase of 61% of courses teaching business ethics in business curricula (Tran 2015). In 2007, the UN launched PRME, an initiative similar to the Global Compact, dedicated to business and management schools.

The United Nations initiative – PRME: a call for change in management education

The Principles for Responsible Management Education (PRME) is a United Nations initiative that aims to promote responsible leadership skills among future leaders (PRME 2019). This initiative was launched to encourage business and management educational institutions to equip their students with social responsibility skills and the ability to balance between profit maximization and sustainability goals. Joining PRME is voluntary, and today, PRME counts a total of 772 signatories worldwide and 15 supporting organizations (PRME 2019). Most of the participants are from Europe and the United States, 38% and 23% respectively. Only five participants are Moroccan institutions. By becoming a PRME signatory, higher education institutions commit to embedding the six principles of the PRME framework (see Table 5.3) in their curricula, research agenda, and student engagement. A taskforce composed of deans, directors, university presidents, officials of leading business schools, and academic institutions developed PRME principles to appear in 2007 at a UN Global Compact Leaders' Summit. United Nations Secretary-General Ban Ki-moon said in the closing remarks, "The Principles for Responsible Management Education have the capacity to take the case for universal values and business into classrooms on every continent" (PRME 2019). As part of their commitment, participants must submit a report either annually or bi-annually summarizing their activities and advancement towards the six principles.

Over the years, PRME has become a platform where stakeholders can exchange best practices and share ideas. Through different global fora and summits organized by PRME, different school leaders interact and engage in stimulating conversations and discussions to improve management education (Tavanti & Wilp 2015).

Although PRME is a promising initiative, its impact remains modest for several reasons. First, the number of business school participants is low compared with the total number of business schools worldwide, which is estimated to be 16,000 institutions (PRME 2019). Second, PRME does not validate the reports or verify their content. The absence of a verification mechanism could be a weakness of PRME (Tavanti & Wilp 2015). Today, of the 772 signatories, about 155 are listed on the PRME website as non-communicating signatories, meaning that they did not submit their PRME report and hence are not upholding their commitment.

Today, the concern for business ethics continues from both business and academic perspectives. Some countries and corporates have been more successful than others in the quest to improve business ethical conduct. There is no silver bullet. Efforts must continue at different levels and in different areas to try to eradicate corruption and promote a clean business environment.

Table 5.3 PRME framework principles

- Principle 1 | Purpose: We will develop the capabilities of students to be future generators of sustainable value for business and society at large and to work for an inclusive and sustainable global economy.
- Principle 2 | Values: We will incorporate into our academic activities, curricula, and organizational practices the values of global social responsibility as portrayed in international initiatives such as the United Nations Global Compact.
- Principle 3 | Method: We will create educational frameworks, materials, processes and environments that enable effective learning experiences for responsible leadership.
- Principle 4 | Research: We will engage in conceptual and empirical research that advances our understanding about the role, dynamics, and impact of corporations in the creation of sustainable social, environmental and economic value.
- Principle 5 | Partnership: We will interact with managers of business corporations to extend our knowledge of their challenges in meeting social and environmental responsibilities and to explore jointly effective approaches to meeting these challenges.
- Principle 6 | Dialogue: We will facilitate and support dialog and debate among educators, students, business, government, consumers, media, civil society organizations and other interested groups and stakeholders on critical issues related to global social responsibility and sustainability.

Source: www.unprme.org/about-prme/the-six-principles.php.

Ethics in the Moroccan business environment

Morocco is a developing country located in the northwest corner of the African continent. It is a stable country politically and is considered a gateway to the African continent, given its strategic geographic location. The mostly visible ethical issue facing the Moroccan business environment is corruption in the public and private sectors. The cost of corruption is up to 7% of the country's GDP (Le Desk 2018), and in some sectors, the cost of corruption can represent up to 10% of the cost of production (MAPNews 2018). Corruption ranks as the second most problematic factor for doing business in Morocco (World Economic Forum 2014). Additionally, according to a survey conducted by Transparency International (2016), more than 30% of the business respondents reported that their company lost new business because a competitor paid a bribe. Hence our discussion of business ethics in the Moroccan business environment will be from the corruption perspective.

The Moroccan business environment, like in many other African countries, is plagued with corruption and unethical business conduct. Given the seriousness of the issue in the African continent, in 2018, the theme of the African Union summit was dedicated to corruption with the title: "Winning the Fight against Corruption: A sustainable path for Africa's transformation" (Songwe 2018).

According to Transparency International (TI) (2018), northern African countries are still facing problems in the fight against corruption. TI uses a tool called the Corruption Perception Index (CPI), which surveys experts and business people to determine countries' perceived levels of corruption, using a scale of 0 to 100, where 0 corresponds to highly corrupt and 100 indicates very clean (Transparency International 2019). According to the CPI of 2018, the least corrupt country was Denmark with a score of 88, while Somalia was the most corrupt country with a score of 10. Morocco's score, however, was 43, which is around the average. Table 5.4 lists Morocco's global ranking and Corruption Perception Index (CPI) for the last five years. Even with the fact that the CPI has improved by 3 points in 2018 compared with 2017, the reality remains that unethical business conduct such as corruption and fraud is hindering the development of business (MAPNews 2018); hence more efforts need to be made in order to foster a clean business environment. Our thesis is that to have ethical managers, we need fresh graduates who are adequately prepared to tackle the ethical challenges of today's work environment, hence the importance of this study in assessing the current status of business ethics education in public and private Moroccan business schools and universities. Business ethics training is a necessary step towards a clean and uncorrupt business environment.

In its efforts to fight business fraud and corruption, the Moroccan government put in place several reforms at different levels. Following the signature of the United Nations Convention against Corruption in 2003 and its ratification in 2007, Morocco established the Central Authority for the Prevention of Corruption (Instance Centrale de Prévention de la Corruption (ICPC)). The ICPC's mission is mainly to coordinate, supervise, and ensure the follow-up of the implementation of anti-corruption policies as well as collecting data on matters related to corruption and fraud to facilitate decision-making. Conversely, the ICPC does not have any authority to investigate and prosecute corruption and fraud cases (ICPC.ma 2019). In 2015, the Moroccan government launched a countrywide strategy aimed at fighting corruption and fraud. The strategy aims to reinforce integrity and significantly reduce corruption by 2025 (MAPNews 2018) via targeting five dimensions, namely governance, prevention, repression, communication and awareness, and education and training (see Figure 5.1) (MMSP.gov.ma 2019). Although education and training are part of the strategy, it is worth noting that only a small portion of the budget allocated to the strategy goes to the education and training dimension. In fact, of the total strategy budget of

Table 5.4 Corruption Perception Index and global ranking of Morocco

Year	*2014*	*2015*	*2016*	*2017*	*2018*
CPI	39	36	37	40	43
Global ranking	80	88	90	81	73
Number of countries	174	167	176	180	180

Source: www.transparency.org/cpi2018.

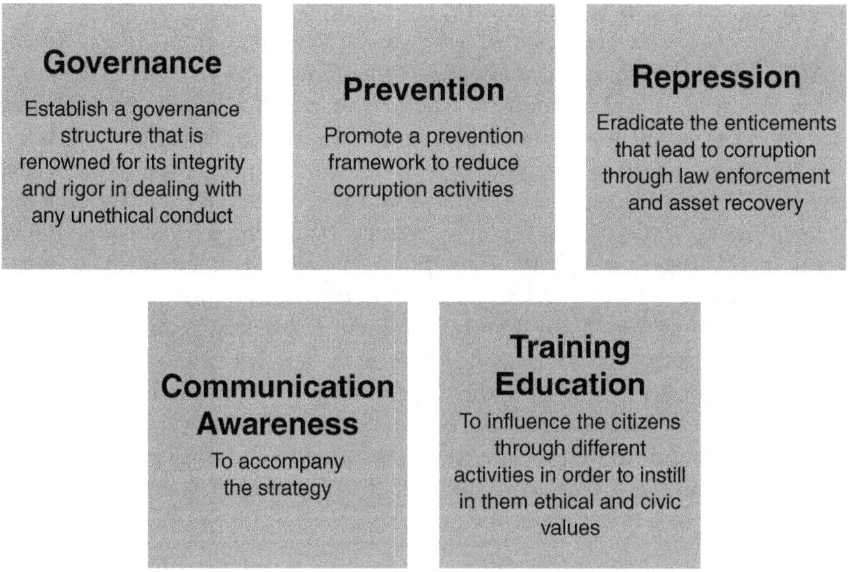

Governance
Establish a governance structure that is renowned for its integrity and rigor in dealing with any unethical conduct

Prevention
Promote a prevention framework to reduce corruption activities

Repression
Eradicate the enticements that lead to corruption through law enforcement and asset recovery

Communication Awareness
To accompany the strategy

Training Education
To influence the citizens through different activities in order to instill in them ethical and civic values

Figure 5.1 Five dimensions of the anti-corruption strategy of Morocco.

Source: www.mmsp.gov.ma/uploads/file/Strategie%20Nationale%20de%20lutte%20Corruption_SNLCC_FR_2016.pdf.

MAD1.796 billion (MMSP.gov.ma 2019), only 1.4% is allocated to education and training, which is unfortunate given the importance of education and training as prevention methods in the fight against unethical behavior (Spector 2012; Makinwa 2013; Indawati, 2015; Baxter et al. 2017; Kamil 2018; Suyantiningsih & Rahmadonna 2019; Stachowicz-Stanusch & Amann 2019; Hauser 2019).

As part of the reform, in 2015, the Ministry of Justice launched an anonymous whistleblowing, toll-free hotline for citizens. In the beginning of the launch, the hotline received an average of 500 calls per day (Telquel.ma 2018). After two and a half years, there were only 36 prosecutions that led to arrests, leading to a broad perception of the failure of Morocco's national strategy (Telquel.ma 2018; Medias24 2016). In 2018, the Prosecutor's Office launched another anonymous whistleblowing, toll-free hotline. This time, in three days, 599 calls were received, and within 48 hours, three civil workers were arrested for corruption (La Nouvelle Tribune 2018). The number of calls received in such a short period demonstrates that Moroccan citizens do not accept corruption, and that bribery is not culturally accepted as is sometimes cited in articles (Boujemi 2012; Berraou 2019)

A significant player in the promotion of business integrity in Morocco is Transparency Maroc. Transparency Maroc is an association created on 6 January 1996 by a group of citizens to address the alarming situation of corruption and lack of transparency and to promote ethics and good governance (Transparency Maroc 2019). It is a non-governmental organization that adheres to the principles of Transparency International (TI), an international non-governmental, non-profit organization based in Berlin whose objective is to fight corruption around the world and prevent criminal activities that result from corruption. Since its creation, Transparency Maroc has been part of the democratic movement working for good governance, the development of citizenship, the promotion of the rule of law, and the establishment of a national integrity system. Following TI's global strategic framework "Together Against Corruption," presented in Figure 5.2, Transparency Maroc's 2017–2020 strategic plan revolves around three elements (Transparency Maroc 2019). The first element is "society," which consists of providing citizens, civil society, and opinion leaders with the necessary tools to identify, condemn, and denounce all forms of corruption and fraud. The second element is the state, which entails using the movement Together Against Corruption to pressure the government and the political community to respect their commitment to the national anti-corruption

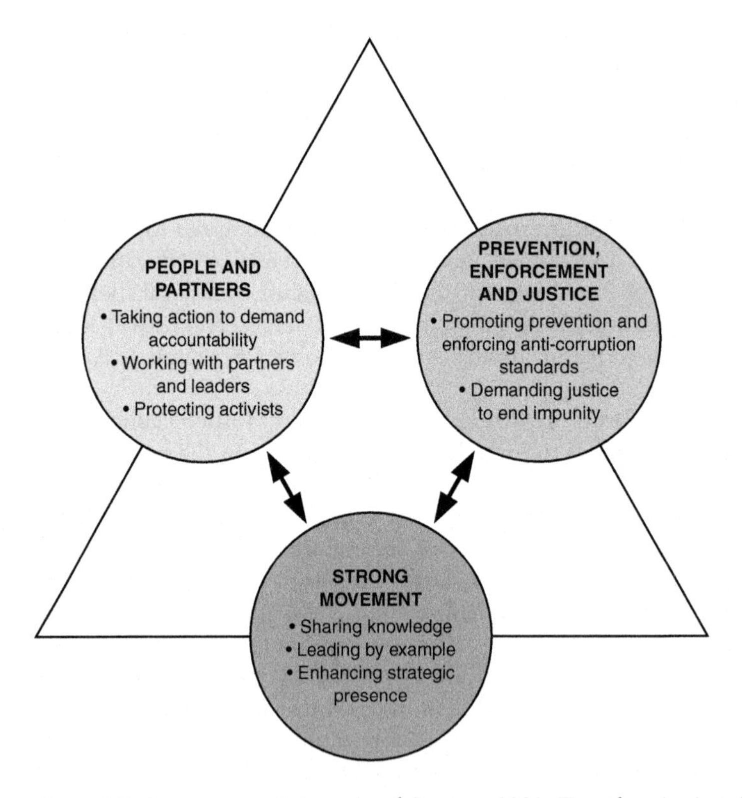

Figure 5.2 Transparency International Strategy 2020: Together Against Corruption.

Source: Together Against Corruption: Transparency International Strategy 2020 (2015) by Transparency International is licensed under CC-BY-ND 4.0.

strategy. The third and last element is the strengthening and engagement of different actors who can take civil society actions against corruption.

Regarding the private sector, the General Confederation of Moroccan Enterprises (La Confédération Générale des Entreprises du Maroc, CGEM), an association with more than 90,000 enterprises in its membership, was created in 1947 to defend the interests of private organizations with public organizations. CGEM is the first representative of the Moroccan private sector and the leading association of businesspeople. In 2007, CGEM announced the creation of a corporate social responsibility certification. This certification is awarded for a three-year period following an evaluation of the candidate company by a third-party expert certified by CGEM. This certification rests on nine principles, listed in Table 5.5. There are about 100 CGEM member companies which have earned the corporate social responsibility certification. These companies benefit from several advantages, including personalized treatment by partner institutions, facilitation of procedures, special pricing, flexibility in control and inspection, and speed in the processing of cases (Benchan 2017).

The above-described efforts are necessary first steps towards a cleaner business environment, but they are not enough. Unfortunately, in Morocco, little is done when it comes to business ethics education. Horn and Kennedy (2008) highlighted the role that business schools have to meet the "social contract with the business community to provide ethical entry-level professionals" (p. 7). Business schools have to play a more prominent role in the promotion of ethical business conduct and adopt different approaches to business ethics education towards better preparing future responsible managers and leaders.

An action-based approach to business ethics education: Giving Voice to Values (GVV)

With the rise of interest in business ethics education in business schools, several scholars have adopted different approaches to teaching students

Table 5.5 Nine principles of the CGEM Corporate Social Responsibility Label

1. Respect human rights
2. Continuously improve working conditions and labor relations
3. Preserve the environment
4. Prevent corruption
5. Respect fair competition rules and regulations
6. Reinforce corporate governance transparency
7. Respect the interest of clients and consumers
8. Promote the social responsibility of suppliers and subcontractors
9. Develop commitment to the community

Source: http://rse.cgem.ma/charte-rse.php.

ethical decision-making. Roberts and Roper (2015) explain four ethical frameworks commonly used in the classroom, namely Kohlberg's stages of moral reasoning, a guided-question-based approach, the use of business codes of ethics, and normative philosophies. Although such approaches help students better comprehend ethical decision-making (Bishop 1992; Oddo 1997; Park 1998; Cornelius et al. 2007; Roberts & Roper 2015), they stand short of showing students how to apply them when dealing with ethical situations (Melé 2008). Hence, some scholars, including Moberg (2006) and Gentile (2010), suggest new ways to enable learners to implement the right course of action when faced with ethical dilemmas. Given the growing popularity of Gentile's approach, Giving Voice to Values (GVV), it is deemed essential to delve into this action-based approach to business ethics education.

The GVV curriculum offers a new approach to teaching business ethics. Whereas traditional approaches to business ethics education deal with the awareness, understanding of ethical dilemmas, and analysis of ethical decision-making, this new approach focuses on the action. GVV asks the question, "once you know what you believe is right, how do you get it done, effectively?" (Arce & Gentile 2015). The GVV premise rests on the fact that in most cases when one is faced with an ethical dilemma, the issue is not what to do, but rather how to act on one's values. As clearly stated by Gentile, "GVV is not about deciding what is the right thing to do, … but about how to get it done" (Gentile 2010). Based on seven pillars, GVV equips individuals with strategies to voice their values and "provides training to build the muscle to do so" (Gentile 2010). These seven pillars are described in Table 5.6 (Gentile 2010, p. 244).

What is unique about GVV is that it emphasizes competency. GVV assumes that many of the people who behave unethically do so because they do not have the competence to voice their values (Freeman et al. 2009). Based on 12 assumptions, summarized in Table 5.7, GVV attempts to evoke "moral competence" for action by empowering the individual to train the muscle of acting on one's values through scripting and rehearsal (Gentile 2010).

Table 5.6 Seven pillars of Giving Voice to Values

1. Values: Recognize that certain values are widely shared
2. Power of choice: Acknowledge that you have a choice
3. Normalization: Treat values conflicts as normal
4. Purpose: Consider your personal and professional purpose
5. Self-knowledge, self-image and alignment: Play to personal strengths
6. Voice: Find your unique way (voice) to express your values
7. Reasons and rationalization: Anticipate rationalizations for unethical behavior

Source: Reprinted from Gentile (2010, p. 244).

Table 5.7 The 12 assumptions of Giving Voice to Values

1. I want to voice and act upon my values.
2. I have voiced my values at some point in my past.
3. I can voice my values more often and more effectively.
4. It is easier for me to voice my values in some contexts than others.
5. I am more likely to voice my values if I have practiced how to respond to frequently encountered conflicts.
6. My example is powerful.
7. Although mastering and delivering responses to frequently heard rationalizations can empower others who share my views to act, I cannot assume I know who those folks will be.
8. The better I know myself, the more I can prepare to play to my strengths and, when necessary, protect myself from my weaknesses.
9. I am not alone.
10. Although I may not always succeed, voicing and acting on my values is worth doing.
11. Voicing my values leads to better decisions.
12. The more I believe it's possible to voice and act on my values, the more likely I will do so.

Source: Adapted from Gentile (2010, pp. 3–23).

The innovative GVV curriculum, developed by former Harvard Business School professor Mary C. Gentile, is increasingly adopted by different universities and corporations alike worldwide. GVV has been piloted in over 1,065 schools, companies, and other organizations worldwide (Givingvoicetovaluesthebook.com 2019) and adopted by more than 125 institutions (thecasecentre.org 2019). An online version of the GVV course is currently available on Coursera and so far over 4,400 people worldwide have enrolled in the course. Additionally, the GVV curriculum, case studies, and teaching plans are available online for educators to use for free. The material is accessible through the Darden Publishing website (store.darden.virginia.edu 2019). The GVV curriculum has been translated into Arabic by the Al Akhawayn University Center for Business Ethics (CBE). Similarly, the GVV book and some material from the curriculum was translated into Mandarin, Korean, and Russian (SSIR.org 2019).

Despite the growing adoption, GVV has been criticized for not being a comprehensive and holistic approach (Gonzalez-Padron et al. 2012). Gonzalez-Padron et al. (2012) argue that GVV underplays the importance of ethical analysis in its approach. Ethical analysis is a tool that consists of understanding the current ethical issue and identifying the risks and implications of decisions before acting. Gonzalez-Padron et al. (2012) maintain that GVV should involve ethical analysis for it to be more effective, since ethical analysis allows the individual to recognize whether the stance adopted is wrong. Nevertheless, Arce and Gentile (2015) explain that "the process of systematically working through the GVV methodology for crafting scripts and action plans tends to emphasize again the necessary ethical analysis and to surface any potential flaws in the assumed ethical position" (p. 538). In fact, GVV does acknowledge the importance of analysis, but goes beyond that by focusing more on ethical implementation (www.darden.virginia.edu/ibis/gvv).

Business ethics education in Morocco: a descriptive, exploratory study

To assess the extent to which business schools in Morocco are embracing business-ethics-related concepts in their missions and programs, an exploratory study and a descriptive case study were conducted. The exploratory study aims to identify whether business ethics courses are offered within higher education institutions and universities, whether the course is mandatory or optional, and whether the course offering is at the undergraduate or graduate level. It also served to see if these institutions offer concentrations and/or minors or full programs in concepts such as CSR, corporate governance, social entrepreneurship, or any other related topic. The descriptive case study aims to present how a business school has integrated business ethics in its business curriculum and beyond. In the context of the study, business ethics is defined in the broader sense to include concepts of corporate social responsibility, sustainability, diversity, corporate governance, environmental issues, social entrepreneurship, and compliance. To achieve the stated goals, both primary and secondary data were used. Primary data was collected using interviews with deans, directors, or program coordinators, and secondary data was collected from the institutions' websites. The target population consists of universities that grant business degrees and schools of business administration in Morocco.

The higher education landscape in Morocco consists of four types of institution: public universities and institutions, private universities and institutions, public professional schools (such as engineering schools), and institutions created by partnerships. In the public sector, management education is offered at schools of law, economics, and social sciences (FSJES) as well as business schools affiliated with universities such as the National Schools of Commerce and Management (les Ecoles Nationales de Commerce et de Gestion, ENCG). The FSJES are accessible to all students with a high school degree, whereas the ENCGs have admission eligibility requirements and tests to earn access. The admission tests explain the difference in the average number of students per type of institution: for the academic year 2017–2018, the average number of students per FSJES faculty was around 20,000, whereas in the ENCGs, it was around 1,400 students. In private schools, the average number of students per school was around 345 students. The 29 public business schools and institutions,

83 private schools and universities, and one public institution created by a partnership grant business degrees at the bachelor and/or master level (ENSSUP.gov.ma 2019).

To determine whether these institutions give any importance to ethics, the published mission statements on their websites were analyzed. Davis et al. (2007) found that students at universities incorporating ethics in their mission statements had significantly higher perceived character trait importance and character reinforcement than those at universities with missions lacking ethical statements. They concluded that "that schools that explicitly stated ethical content in their mission statements do influence student ethical orientation" (Davis et al. 2007, p. 11) Additionally, accrediting agencies also highlight the importance of the mission statement for business higher education institutions. For instance, the AACSB clearly states in its 2013 standards that "a quality business school has a clear mission, acts on that mission, translates that mission into expected outcomes, and develops strategies for achieving those outcomes" (AACSB 2018, p. 15). Similarly, the EFMD highlights in its EQUIS accreditation standards that "The School should have a clearly articulated mission that is understood and shared throughout the institution" (EFMDglobal.org 2019, p. 8).

All 113 websites were surveyed to review the schools' mission statements. Fifteen private schools did not have a website or their website was not accessible. For the remaining 98 schools, if the mission statement is not available, the message of the dean/director or the institution's objectives were reviewed to see if there is any mention of ethics or related topics. Table 5.8 summarizes the findings separated by type of school. Twenty-five of the 30 public institutions do not have a mission statement, while four institutions have mission statements but do not mention ethics or CSR or any related topic. Only one public business school has a mission statement directly mentioning ethics. As for the private institutions, only 18 schools have an explicitly stated mission statement posted on their websites, and 12 of them mention ethics. To summarize, 16% of the public schools mention ethics in either mission statements, the messages of the dean, or objectives compared with 20% in the private sector. Only 21% of all business schools with available websites mention ethics or related concepts in their mission statements, the messages of the dean, or objectives. These findings reveal the dire need for a lot more to be done.

Next, selected deans, directors, and administrators were interviewed to collect data on the extent to which their schools offer ethics-related courses, program offerings, and/or co-curricular activities. This information is not available on the websites, as most schools do not publish their academic catalogs online and present only general information about the programs. Therefore, structured interviews of a small sample of deans, directors, and administrators were conducted. Using convenience sampling, 20 deans, directors, and administrators were first solicited by phone or via

Table 5.8 Ethics mentioned in missions or dean's message

Public institutions	30
Have mission statement	5
Ethics mentioned in mission statement	1
Ethics mentioned in message of the dean or objectives	4
Ethics not mentioned in message of the dean or objectives	23
No website	0
Private institutions	83
Have mission statement	18
Ethics mentioned in mission statement	12
Ethics mentioned in message of the dean or objectives	4
Ethics not mentioned in message of the dean or objectives	43
No website	15

Source: institutions' websites.

email. Those who have responded were later contacted to schedule the interview. Given the nature of some of the questions requiring a search for information within the institutions, some respondents preferred to provide answers to some questions via email. For incomplete questionnaires, follow-up phone calls were conducted to clarify some information when needed.

Of the 20 schools contacted, eight were public and 12 were private. The respondents consisted of five public schools and eight private schools, representing an overall response rate of 65%. Of the 13 responding schools, six (four public and two private) did not have any mission statement.

Regarding stand-alone ethics courses, five schools offered mandatory courses at the undergraduate level, and six schools offered courses at the graduate level in different programs. Only one school offered elective courses in ethics in addition to one mandatory course at the undergraduate level. Four schools did not offer any courses on the topic. It is worth noting that among these four schools, three were private schools with mission statements mentioning ethics. When asked about the reason for this discrepancy, they noted that they had opted for an integrative approach and embedded business ethics concepts in different mandatory courses, such as organizational behavior, entrepreneurship, and introductory management. Table 5.9 lists the reported course titles at the undergraduate and graduate levels. Most of the reported course titles were related to concepts of corporate social responsibility, governance, and ethics.

Concerning programs and concentrations, only one public school offered a master's program – in social innovation and management of social solidarity economy. Another public school stated that they are launching a new minor in values-driven leadership in the next academic year.

Table 5.9 Surveyed business schools' ethics-related course titles and counts

Undergraduate courses	*Count*
Geostrategy and sustainable development	1
Corporate social responsibility	3
Business ethics in Islam (Ethique des Affairs en Islam)	1
Business law and ethics	1
Ethics in information technology	1
Mindfulness	1
Graduate courses	
Ethics of organizations (Ethique des Organisations)	1
Corporate social responsibility	5
Social innovation	1
Sustainable development	1
Ethics and management	1
Managing sustainability in emerging markets	1
Business ethics	1
Financial ethics and corporate governance	1
Business ethics and corporate governance	1

Source: interviews with deans, directors and program coordinators.

The exploratory study also reviewed extracurricular contributions to ethics education. Agle et al. (2011) stated that while ethical concepts can be learned and discussed in ethics courses, students can develop moral character and leadership through extracurricular activities such as community service, study abroad, volunteering, and leading organizations within their universities or schools. Accordingly, respondents were asked to describe any activities undertaken by students outside the classroom that expose them to concepts of business ethics. The most cited activity is guest speakers presenting topics such as CSR, compliance, and social entrepreneurship. These talks are either organized by schools or at times by student clubs. Respondents recognized the lack of focus on ethics issues; even when schools adopt missions highlighting ethics, CSR, and sustainability, or offer courses in ethics, they do not complement these efforts with co-curricular activities to further address ethics issues. Only one school has a stand-alone center dedicated to business ethics. This center organizes awareness campaigns, workshops, and poster exhibitions as part of its activities. It is worth noting that this school stands out from the pool of respondents as one that does more than the others when it comes to mainstreaming business ethics education in its activities. This school will be highlighted in the case study below as a practical example of integrating and mainstreaming business ethics in its programs.

The case of Al Akhawayn University in the Ifrane School of Business Administration

Al Akhawayn University (AUI) is a Moroccan public, not-for-profit, liberal arts university. In 2017, AUI earned the accreditation of the New England Commission of Higher Education (NECHE). It consists of three schools: School of Business Administration (SBA), School of Science and Engineering (SSE), and School of Humanities and Social Sciences (SHSS). SBA is the largest school within the university, with over 50% of the student body. It presents a model for how a business school can embrace business ethics education, as evidenced in its different dimensions through which business ethics permeates the school culture among its students, faculty, staff, and alumni.

Disclosure: both authors are affiliated with the university featured in the case study. The sections below outline some of the initiatives, launched within the university and the school, to inculcate business ethics values in students.

Mission and values

The Al Akhawayn SBA mission is "To shape future ethical, successful managers and leaders with a local and global perspective." The importance of this statement stems from the participatory approach used to define it. The process involved all stakeholders, including students, alumni, faculty, staff, and administrators, who gathered in a retreat and drafted the mission. The mission has become part of the culture of the school and the glue that binds all stakeholders together. It is purposefully placed at the entrance of the school, and instructors are encouraged to remind students of the mission in their classes.

The liberal arts curriculum

Unlike most business schools in Morocco, SBA adopts a liberal arts curriculum that exposes students to subjects beyond the specific business topics. Subjects such as science, social sciences, and humanities broaden the perspectives of a business student. About 50% of the SBA curriculum consists of such subjects as philosophical thought, art and architecture, sociology, biology, and many other choices from the humanities and social sciences. Janeksela (2012) finds that liberal arts education teaches students valuable skills, such as ethical reasoning skills. Additionally, Hansen (2004) reports that employers describe liberal arts graduates as "well rounded," having the ability to see the "big picture." The underlying rationale for SBA adoption of the liberal arts approach is to nurture "critical thinking," stimulate "inquiring minds," and champion "an awareness of the perennial questions and new challenges that confront humanity, a depth and consistency of moral judgment, the ability to speak and write with clarity and precision, a capacity and life-long desire for learning, the exchange of ideas and knowledge for development, and an awareness of other cultures beyond national borders" (AUI 2017–2019 Catalog 2017, p. 119). SBA embraces a learning environment wherein future managers explore different streams of knowledge towards questioning underlying beliefs, examining perspectives, and building a rounded learning experience.

Additionally, the business core courses include a compulsory course in business law and ethics. Furthermore, ethical concepts are integrated into several courses where faculty cover ethical issues within their specific disciplines. In fact, out of 15 core business courses, 10 directly address business ethics in their intended learning outcomes. The appendix includes a sample syllabus as well as the bachelor of business administration curriculum map highlighting the courses addressing the ethics program learning outcome.

Community involvement program

Completing coursework is not enough for graduation at SBA; instead, business students are required to fulfill the requirements of the Community Involvement Program (CIP). Commonly known as social internship, students attend seminars on human development and the role civil society plays. Then, they have to consult a database of over 330 civil society organizations, both in Morocco and abroad, to pick an NGO or charity association to carry out voluntary work for a minimum of 60 hours. Once their placement is approved, students have the choice to space out the volunteering work during an academic semester, thus they focus their volunteering on the local community where the university is based (AUI Office of Community Involvement n.d.). Many of these students opt to volunteer for the diverse activities of Al Akhawayn's Azrou Center for Community Development, which organizes medical campaigns and offers subsidized programs in schooling, capacity-building, and healthcare to the local under-served population of 3,000 beneficiaries (AUI Social Responsibility n.d.). Alternatively, they dedicate their inter-semester breaks for volunteering in their selected organizations usually based in their hometowns. Once the volunteering work is completed, students get together for "Experience Sharing Roundtables," elaborate their reports to reflect on their lessons learned, and present their community work to a jury of faculty and staff.

Making community involvement an undergraduate requirement towards graduation is undeniably a significant step towards tackling the research findings on business students' disturbing attitudes (Friedman 1970; Marwell & Ames 1981; Grant 1991; Frank et al. 1993; Melé 2008) geared towards "stigmatizing" those in need of help (Giacalone & Promislo 2013).

The Leadership Development Institute (LDI)

Further to the undergraduate community involvement requirement, business students have the option to build skills to lead by joining the Leadership Development Institute (LDI). These are "high-achieving, highly motivated students" who secure "the opportunity to sharpen their skills and earn a co-curricular certificate" (AUI 2013–2015 Catalog 2013, p. 85). The LDI program supports existing courses by working towards fulfilling the mission to "develop leaders that serve" by combining "transformational, servant, social change and other models of leadership with best practices in student development and an active learning process of social engagement with a global perspective" (LDIIfrane.org n.d.). Integrity, ethics, innovation, and "bottom-up approach to leadership – leadership does not come with position" are some of the LDI guiding principles (AUI 2017–2019 Catalog 2017, p. 116). The LDI has developed an annual tradition, for the past eight years, of raising funds to award the three most impactful civil society organizations within the metropolitan area where the university is based: Fez-Meknes region (FesNews.net 2018).

Extracurricular activities

Beyond curricular requirements, SBA future managers are actively encouraged to engage in extracurricular activities. With over 40 active student-led clubs, SBA students have a wide range of options to pursue experiential learning (AUI Student Organizations n.d.). Students usually like to experiment with creating and leading clubs, but a large number flock around "socially responsible student organizations." These include Hand in Hand, Rotaract Club, and The Islamic Arts and Culture, which raise significant funds (MAD630,000, MAD370,000, and MAD10,000 respectively) during their annual events to carry out community projects, including school renovations, boarding dorms' set-up, and equipment for women income-generation activities (AUI Community Involvement n.d.). For instance, AUI Rotaract Club, which strives to "foster leadership and responsible citizenship, encourage high ethical standards in business" won "Best Action for Public Interest" at the conference held in Tunisia in March 2011, for Rotaract Middle East and North Africa (MENA) District (AUI Rotaract Club n.d.). Founded in 1995, the student-run Hand in Hand association has led a wide range of community development initiatives to tackle issues of school attrition, medical coverage, and winter heating (AUI Hand in Hand n.d.).

SBA signatory of PRME

In 2013, the School of Business Administration was the first school within a Moroccan public university to become a signatory of the Principles of Responsible Management Education (PRME 2019). Since then, the school submitted two reports highlighting the different activities related to the six principles of PRME. These reports are available on the PRME website (PRME 2019). As a signatory, SBA commits to integrating PRME principles in its curriculum. Faculty are encouraged to embed these principles in their courses and research.

The Center for Business Ethics

In 2015, AUI inaugurated the Center for Business Ethics, funded by Siemens under the Siemens Integrity Initiative. The center's mission is to promote integrity and business ethics, both in schools and the private sector. It achieves its mission through awareness campaigns, training of students, faculty, and professionals, research, and curriculum development. Since its launch, the center has organized workshops designed to train faculty on business ethics concepts. Topics of the workshops include Giving Voice to Values, business ethics case study writing, and teaching business ethics. Over 50 professors and instructors from 13 different public and private institutions benefited from these workshops. Additionally, the center organized awareness campaigns, including guest lectures, conferences, and workshops targeting both undergraduate and graduate students, reaching over 1,200 students in different private and public schools and universities nationwide. In 2018, the center organized a national business ethics teaching case study competition, which resulted in the selection of seven case studies covering real, local ethical issues. These cases will be available for free on the center's website.

Giving Voice to Values at SBA

The Center for Business Ethics (CBE) sparked interest among business educators, students, and professionals alike when it featured the new business ethics approach, Giving Voice to Values (GVV). Many CBE participants expressed ease with the GVV action-based approach to business ethics, as it acknowledges that students/users already have an intrinsic "moral compass," operating in varying degrees in determining moral judgments in the face of ethical dilemmas (Edwards & Kirkham 2014). Discussions with CBE participants revealed the new GVV approach lessens the burden on emphasizing morality, which is found to be one source of intimidation for business educators (Roberts & Roper 2015).

Taking note of the increasing interest in GVV, the CBE carried out various initiatives. GVV founder, Professor Mary Gentile, was featured in several lectures addressing business educators, students, and professionals nationwide. She also had roundtable discussions with students aspiring for leadership building at LDI. Two SBA professors championed piloting the GVV curriculum in their classes and developed local GVV cases.

The above-described dimensions demonstrate how the Al Akhawayn School of Business Administration (SBA) commits to its mission of educating future managers and leaders with strong values and a high sense of responsibility. Through these dimensions, the SBA attempts to respond to the growing demand for values-driven leadership and a clean business environment. The SBA presents an excellent model that can be followed by other business schools in Morocco and the region.

The findings of this study reveal that business ethics education in Morocco is still in its infancy. Business ethics remains superficially addressed compared with other developed countries. Private schools tend to give the topic more importance than public schools, which are home to the majority of students, with the missed opportunity to learn and internalize business ethics. Even if the sample size is small, the composition of the respondents is varied enough to give a good picture of the current situation. The case study could serve as a role model for other institutions on how to integrate business ethics inside and outside of the classroom. Further research is needed to cover a larger sample of schools and address different questions. Additionally, surveying faculty and student perspectives on business ethics education would shed light on the factors hindering its widespread adoption in the business curriculum. The role of the Ministry of Higher Education is equally crucial in making business ethics courses a requirement for a program to be accredited. Future research may also explore the content of the courses offered.

Conclusion

The purpose of this chapter is to highlight the importance of business ethics education in the fight against corruption and to capture the current status of business ethics education in Moroccan business schools. It argues that in order to promote clean business and enhance business integrity, a focus on business ethics education is a necessary condition. The exploratory, descriptive study reveals that business ethics education in Morocco is still in the early stage of development. Private schools seem to give the topic a little more importance than public schools do. While there are several efforts to eradicate unethical business behavior in Morocco, several challenges remain. Education should be at the heart of these efforts for better results to be garnered, which requires a strong commitment from higher education leaders and scholars of many institutions. Starting from mission statements to curricula and extracurricular activities, business schools' leaders have the opportunity to make a difference and facilitate lasting impacts within their institutions and beyond. In order to improve the current situation, it is recommended that the Ministry of Higher Education requires all business schools to include a stand-alone course on business ethics as part of its accreditation process, which would constitute a significant step toward producing future generations of managers who are better equipped to handle the complex challenges of the workplace.

The chapter also presents different approaches to business ethics education adopted by business schools worldwide, namely PRME and GVV, and argues that these two initiatives provide promising strategic directions for business schools.

The practical case of the Al Akhawayn School of Business Administration demonstrates that business ethics education is a multidimensional concept that cannot be limited to offering stand-alone courses and using an integrative approach. Instead, it can be included in all activities inside and outside of the classroom to produce graduates who "enhance Morocco and engage the world" (AUI 2019).

Appendix

Sample Syllabus

MIS 3399 – Ethics in Information Technology

School of Business Administration

I General information

Course Number:	MIS 3399	Course Title:	Ethics in Information Technology
Credits:	3 SCH	Prerequisites:	MIS 3301

II Course description

This course introduces students to the ethical issues surrounding the use, development and management of information technology. As IT invades our personal and professional lives, students need to understand the implications of using IT as well as their rights and obligations as users, managers, and/or marketer of IT products. In this course, we will discuss ethical issues ranging from privacy, security, intellectual property, discrimination, to policies and codes of ethics. Through real cases, we will analyze situations from individual and organizational perspectives. Additionally, we will discuss the ethical issues related to specific IT such as big data, data mining, etc....

III Required textbook

Ethics in Information Technology 6th Edition, by George W. Reynolds

IV Intended learning objectives

After taking this course, students will be able to:

- Discuss the ethical issues surrounding the use of information technology and the challenges it presents locally and globally.

- Critically assess the ethical dilemmas and identify feasible courses of action.
- Develop scripts for ethical situations they may face in the future using the Giving Voice to Values framework.
- Communicate effectively in writing and orally.
- Demonstrate effective teamwork abilities in the course's project.

V Course ILOs mapping to program ILOs

Table A

	BBA Program Learning Goals/ Objectives	*Course ILO Metric (What)*	*Course Assessment Metric (How)*
1. Ethical: Make decisions ethically			
a	Understand ethical concepts including sustainability and concepts of corporate social responsibility [Understanding].	Discuss the ethical issues surrounding the use of information technology and the challenges it presents locally and globally.	Assessed through specific exam questions.
b	Analyze ethical situations [Remembering, Analyzing].	Critically assess the ethical dilemmas and identify feasible courses of action.	Assessed through case studies and course project.
c	Make good judgments (evaluate) in business situations from an ethical perspective [Remembering, Evaluating].	Develop scripts for ethical situations they may face in the future using the GVV framework.	Assessed through class and take-home assignments.
2. Successful: Display skills and competencies of successful business decision-maker			
d	Communicate effectively orally and in writing [Creating].	Communicate effectively in writing and orally.	Assessed through a combination of course project report and presentation.
e	Work effectively within a team [Applying, Analyzing, Evaluating, Creating].	Demonstrate effective teamwork abilities in the course's project.	Assessed through peer evaluation and the presentation part of the course's project.
3. Local: Understand the Moroccan business environment			
c	Assess current business strategies of Moroccan firms.	Discuss the ethical issues surrounding the use of information technology and the challenges it presents locally and globally.	Assessed through course project.

	BBA Program Learning Goals/ Objectives	*Course ILO Metric (What)*	*Course Assessment Metric (How)*
4. Global: Comprehend the global business environment			
a	Understand the global business environment and challenges [Understanding].	Discuss the ethical issues surrounding the use of information technology and the challenges it presents locally and globally.	Assessed through (i) course project, (ii) case studies, and (iii) specific exam questions.

VI Grading

General: Grades will be assigned to four types of activities:

Table B

Activity	*Weight*	*Description*
2 exams	40 points	One midterm exam and one final exam.
Quizzes	15 points	Unannounced quizzes.
Projects	30 points	Report (15%), Presentation (10%), Peer Evaluation (5%) Description of the project will be presented in class in due time.
GVV assignments	10 points	GVV exercise – Tale of two stories.GVV case studies.
Class participation	5 points	Participation in in-class discussions. Not just attendance!

VII Course schedule

The course schedule is a dynamic document (which means we can adapt it as we go along). Changes to the course schedule will be announced in class and via Jenzabar.

Table C

Schedule of Events Week	*Topic*	*Assignments, Exams*
Week 1	Introduction to the course Overview of ethics	Read Chapter 1
Week 2	Ethics for IT workers and IT users	Read Chapter 2
Week 3	Computer and internet crime	Read Chapter 3
Week 4	Privacy	Read Chapter 4
Week 5	Introduction to GVV	Read GVV assigned readings
Week 6	Freedom of expression	Read Chapter 5

Continued

Table C continued

Schedule of Events Week	*Topic*	*Assignments, Exams*
Week 7	Review session and midterm exam	
Week 8	Intellectual property	Read Chapter 6
Week 9	Software development	Read Chapter 7
Week 10	Guest speaker	
Week 11	Impact of IT on productivity and quality of life	Read Chapter 8
Week 12	Social networking	Read Chapter 9
Week 13	Ethics of IT organizations	Read Chapter 10
Week 14	Project presentations & discussion	
Week 15	Project presentations & discussion	
Week 16	Final exam	Final exam will occur on the date and time provided by the university.

VIII Methods of instruction

The case study method will be the main method of instruction. Concepts will be illustrated through case studies and real-life examples and discussed in class. Some concepts will be introduced through lectures. Students' participation will be encouraged through different formats, including individual and group presentations and small and large group discussions. An experiential learning project will be included where students will choose a company and conduct an ethics IT audit.

IX Participation

Class participation is key to your learning process. Through active participation you will learn to express your ideas and demonstrate and enhance your understanding of the different concepts and topics discussed in class. Engaging is meaningful discussions, asking relevant questions, contributing ideas, commenting on the readings, and sharing information are all acceptable participation methods.

X Attendance policy

I will strictly adhere to the university attendance policy, which states that "Attendance is mandatory; you are allowed only 5 unexcused absences during the semester. Then you will be dropped from the course with a failing grade on the 6th unexcused absence."

XI Make-up exam policy

No make-up exams/quizzes will be given. All exams/quizzes must be taken at the scheduled time on the scheduled date. NO EXCEPTIONS!

Bachelor of Business Administration Curriculum Mapping Matrix

Table D Program learning outcomes related to ethics and the courses that cover it

Learning Goals/Objectives	*ACC 2301*	*FM 3301*	*MGT 3301*	*MKT 3301*	*GBU 3302*	*GBU 3303*	*MGT 3302*	*MGT 3303*	*MRS 3301*	*MGT 4301*
Ethics: Make decisions ethically										
a. Demonstrate understanding of ethnical concepts including sustainability and corporate social responsibility	x	x	x	x		x		x	x	x
b. Analyze ethical situations								x	x	
c. Evaluate business situations from an ethical perspective					x		x		x	

Table E

Course Code	*Course Title*
ACC 2301	Accounting Principles I
FIN 3301	Principles of Finance
MGT 3301	Principles of Management
MKT 3301	Principles of Marketing
GBU 3302	Business Law and Ethics
GBU 3203	Enterprises, Markets, and the Moroccan Economy
MGT 3302	Entrepreneurship
MGT 4303	Operations Management
MIS 3301	Management Information Systems
MGT 4301	Capstone Course: Business Policy and Corporate Strategy

XII Cheating policy

Cheating of any kind is NOT tolerated (this includes plagiarism). Cheating will result in an automatic F for the whole course.

XIII Late-work policy

No assignment is accepted after the set deadline.

XIV Available resources for all students

Several services are available to you at the university. I encourage you to use these services when needed to improve your academic experience at AUI.

Tutoring service, writing center, language help (see catalog pages 83–84), counselling services (see catalog page 77), and other student support such as Center for Learning Excellence and Counseling (see catalog page 77).

References

AACSB.edu (2018) '2013 Eligibility Procedures and Accreditation Standards for Business Accreditation' [online] Available at: www.aacsb.edu/-/media/aacsb/docs/accreditation/business/standards-and-tables/2018-business-standards.ashx?la=en&hash=B9AF18F3FA0DF19B352B605CBCE17959E32445D9 (Accessed 14 July 2019).

Abend, G. (2013) 'The Origins of Business Ethics in American Universities, 1902–1936,' *Business Ethics Quarterly*, 23(2), pp. 171–205. doi: 10.5840/beq201323214. http://search.ebscohost.com/login.aspx?direct=true&db=e6h&AN=87352405&site=eds-live.

Abend, G. (2014) *The Moral Background: An inquiry into the history of business ethics*. Princeton: Princeton University Press. Available at: http://search.ebscohost.com/login.aspx?direct=true&db=nlebk&AN=667981&site=eds-live (Accessed 14 July 2019).

Abend, G. (2016) 'How to Tell the History of Business Ethics,' *Zeitschrift fuer Wirtschafts- und Unternehmensethik*, 17(1), p. 42. Available at: http://search.ebscohost.com/login.aspx?direct=true&db=edb&AN=114760797&site=eds-live (Accessed 14 July 2019).

Adler, P. S. (2002) 'Corporate Scandals: It's time for reflection in business schools,' *Academy of Management Executive*, 16(3), p. 148. Available at: http://search.ebscohost.com/login.aspx?direct=true&db=edb&AN=8540425&site=eds-live (Accessed 14 July 2019).

Agle, B. R., Thompson, J., Hart, D., Wadsworth, L., & Miller, A. (2011) Meeting the Objectives of Business Ethics Education: The Marriott School Model and Agenda for Utilizing the Complete Collegiate Educational Experience. In C. Wankel & A. Stachowicz-Stanusch (eds.) *Management Education for Integrity*. Bingley, UK: Emerald Group Publishing Limited, pp. 217–242.

Arce, D. G., & Gentile, M. C. (2015) 'Giving Voice to Values as a Leverage Point in Business Ethics Education,' *Journal of Business Ethics*, 1(31), pp. 535–542.

AUI (2019) History and Mission. [Online] Available at: www.aui.ma/en/about/general/history-mission.html.

AUI 2013–2015 Catalog (2013) [online] Available at: www.aui.ma/academic-catalog/2013-2015.pdf (Accessed 9 June 2019).

AUI 2017–2019 Catalog (2017) [online] Available at: www.aui.ma/images/catalog_2017-2019.pdf (Accessed 7 June 2019).

AUI Community Involvement (n.d.) [online] Available at: www.aui.ma/images/AUI%20and%20GUNI%20Report%202013.pdf (Accessed 9 June 2019).

AUI Hand in Hand (n.d.). [online] Available at: www.aui.ma/en/hih.html (Accessed 9 June 2019).

AUI Office of Community Involvement (n.d.). [online] Aui.ma. Available at: www.aui.ma/en/communityinvolvement.html (Accessed 14 July 2019).

AUI Rotaract Club (n.d.). [online] Aui.ma. Available at: www.aui.ma/en/civic-engagement/student-solidarity-associations/rotaract-club.html (Accessed 9 June 2019).

AUI Social Responsibility (n.d.) [online] Aui.ma. Available at: www.aui.ma/en/civic-engagement/social-responsibility/overview.html (Accessed 9 June 2019).

AUI Student Organizations (n.d.). [online] Available at: www.aui.ma/en/activities/student-organizations/sao-organizations.html (Accessed 9 June 2019).

Baxter, R. J., Holderness Jr., D. K., & Wood, D. A. (2017) 'The Effects of Gamification on Corporate Compliance Training: A Partial Replication and Field Study of True Office Anti-Corruption Training Programs,' *Journal of Forensic Accounting Research*, 2(1), p. A20. Available at: http://search.ebscohost.com/login.aspx?direct=true&db=edb&AN=127822475&site=eds-live (Accessed 8 November 2019).

Benchan, I. (2017) 'La CGEM, précurseur en matière d'édition de normes RSE' [online] Lavieeco. Available at: www.lavieeco.com/dossiers-speciaux/la-cgem-precurseur-en-matiere-dedition-de-normes-rse/ (Accessed 14 July 2019).

Bennis, W. G., & O'Toole, J. (2005). 'How Business Schools Lost Their Way,' *Harvard Business Review*, (5), 96.

Berraou, M. (2019) The Limits of Anti-Corruption Policies in Morocco. [Online] Available at: https://mipa.institute/6551.

Bird, F. (2016) 'Learning from History,' *Zeitschrift fuer Wirtschafts- und Unternehmensethik*, 17(1), p. 8. Available at: http://search.ebscohost.com/login.aspx?direct=true&db=edb&AN=114760795&site=eds-live (Accessed 14 July 2019).

Bishop, T. R. (1992) 'Integrating Business Ethics into an Undergraduate Curriculum,' *Journal of Business Ethics*, 11(4), p. 291.

Boujemi, H. (2012) The Internet and Active Citizenship. Global Information Society Watch 2012: The Internet and Corruption, pp. 172–175.

Buff, C., & Yonkers, V. (2004). 'How Will They Know Right from Wrong? A Study of Ethics in the Mission Statements and Curriculum of AACSB Undergraduate Marketing Programs,' *Marketing Education Review*, 14(3), pp. 71–79.

Cornelius, N., Wallace, J., & Tassabehji, R. (2007). 'An Analysis of Corporate Social Responsibility, Corporate Identity, and Ethics Teaching in Business Schools,' *Journal of Business Ethics*, 76, 117–135. doi: 10.1007/s1055100692716.

Currie, G., Knights, D., & Starkey, K. (2010) 'Introduction: A Post-crisis Critical Reflection on Business Schools,' *British Journal of Management*, p. s1. Available at: http://search.ebscohost.com/login.aspx?direct=true&db=edsbl&AN=RN265428200&site=eds-live (Accessed 14 July 2019).

Davis, J. H., Ruhe, J. A., Lee, M., & Rajadhyaksha, U. (2007) 'Mission Possible: Do School Mission Statements Work?,' *Journal of Business Ethics*, 70 (1), p. 99.

DeGeorge, R.T. (1987) 'The Status of Business Ethics: Past and Future,' *Journal of Business Ethics*, 6(3), pp. 201–211. Available at: http://search.ebscohost.com/login.aspx?direct=true&db=edszbw&AN=EDSZBW259083194&site=eds-live (Accessed 13 July 2019).

DeGeorge, R. T. (1989) 'There is Ethics in Business Ethics; but There's More As Well,' *Journal of Business Ethics*, (5), p. 337. Available at: http://search.ebscohost.com/login.aspx?direct=true&db=edsggo&AN=edsgcl.7864011&site=eds-live (Accessed 13 July 2019).

Edwards, M., & Kirkham, N. (2014) 'Situating "Giving Voice to Values": A Metatheoretical Evaluation of a New Approach to Business Ethics,' *Journal of Business Ethics*, 121(3), pp. 477.

EFMDglobal.org. (2019) 'EFMD Quality Improvement System: The EFMD Accreditation for International Business Schools' [online] Available at: https://efmdglobal.org/wp-content/uploads/EFMD_Global-EQUIS_Standards_and_Criteria.pdf (Accessed 14 July 2019).

ENSSUP.gov.ma (2019) 'Ministère de l'Enseignement Supérieur de la Recherche Scientifique et de la Formation des Cadres' [online] Available at: www.enssup.gov.ma/fr (Accessed 15 July 2019).

FesNews.net (2018) 'Al Akhawayn University Students crown an association that cares for cemeteries in Meknes with "Leader of the Year Award"' [online]. Available at: http://fesnews.net/ (Accessed 9 June 2019).

Frank, R. H., Gilovick, T. D., & Regan, D. T. (1993) 'Does Studying Economics Inhibit Cooperation?,' *Journal of Economic Perspectives*, 7, pp. 159–171.

Freeman, R. E., Stewart, L., & Moriarty, B. (2009) 'Teaching Business Ethics in the Age of Madoff,' *Change: The Magazine of Higher Learning*, 41(6), pp. 37–42. Available at: http://search.ebscohost.com/login.aspx?direct=true&db=edsbl&AN=RN261105194&site=eds-live (Accessed 13 July 2019).

Friedman, M. (1970). 'The Social Responsibility of Business is to Increase its Profits,' *New York Times Magazine*, 30 September.

Gentile, M. C. (2010). *Giving Voice to Values: How to speak your mind when you know what's right*. New Haven, CT: Yale University Press.

Giacalone, R. A., & Promislo, M. D. (2013) 'Broken When Entering: The Stigmatization of Goodness and Business Ethics Education,' *Academy of Management Learning & Education*, 12(1), p. 86. doi: 10.5465/amle.2011.0005A.

Gioia, D. A. (2002) 'Business Education's Role in the Crisis of Corporate Confidence,' *The Academy of Management Executive*, (3), 142.

Givingvoicetovaluesthebook.com (2019) *Giving Voice to Values: How to speak your mind when you know what's right.* [online] Available at: www.givingvoicetovaluesthebook.com/ (Accessed 18 June 2019).

Gonzalez-Padron, T. L., Ferrell, O. C., Ferrell, L., & Smith, I. A. (2012) 'A Critique of Giving Voice to Values Approach to Business Ethics Education,' *Journal of Academic Ethics*, 10, pp. 251–269.

Grant, C. (1991) 'Friedman Fallacies,' *Journal of Business Ethics*, 10(12), 907.

Hansen, K. (2004) Ten ways to market your liberal arts degree. [Online] Available at www.quintcareers.com/marketing_liberal-arts_degrees.html.

Harding, L. (2019) 'Deutsche Bank Faces Action over $20bn Russian Money Laundering Scheme' [online] Available at: https://www.theguardian.com/business/2019/apr/17/deutsche-bank-faces-action-over-20bn-russian-money-laundering-scheme (Accessed 26 March 2020).

Hauser, C. (2019) 'Fighting Against Corruption: Does Anti-corruption Training Make Any Difference?,' *Journal of Business Ethics*, 1, p. 281. Available at: http://search.ebscohost.com/login.aspx?direct=true&db=edsgao&AN=edsgcl.598545779&site=eds-live.

Henderson, V. E. (1988) 'Can Business Ethics Be Taught?,' *Management Review*, 77(8), p. 52.

Holland, D., & Albrecht, C. (2013) 'The Worldwide Academic Field of Business Ethics: Scholars' Perceptions of the Most Important Issues,' *Journal of Business Ethics*, 117(4), p. 777.

Holosko, M. J., Winkel, M., Crandell, C., & Briggs, H. (2015) 'A Content Analysis of Mission Statements of Our Top 50 Schools of Social Work,' *Journal of Social Work Education*, 51(2), p. 222.

Horn, L., & Kennedy, M. (2008) 'Collaboration in Business Schools: A Foundation for Community Success,' *Journal of Academic Ethics*, 6 (1), pp. 7–15.

ICPC.ma. (2019) INPPLC. [online] Available at: www.icpc.ma/wps/portal (Accessed 15 July 2019).

Indawati, N. (2015) 'The Development of Anti-Corruption Education Course for Primary School Teacher Education Students,' *Journal of Education and Practice*, 6(35), pp. 48–54. Available at: http://search.ebscohost.com/login.aspx?direct=true&db=eric&AN=EJ1086370&site=eds-live (Accessed 8 November 2019).

Janeksela, G. M. (2012) 'The Value of a Liberal Arts Education,' *Academic Exchange Quarterly*, 16(4), pp. 1096–1453.

Kamil, D. (2018) 'Fighting Corruption through Education in Indonesia and Hong Kong: Comparisons of Policies, Strategies, and Practices,' *Al-Shajarah: Journal of the International Institute of Islamic Thought & Civilization*, p. 155. Available at: http://search.ebscohost.com/login.aspx?direct=true&db=edb&AN=135414774&site=eds-live (Accessed 8 November 2019).

Khurana, R. (2007). *From Higher Aims to Hired Hands: The Social Transformation of American Business Schools and the Unfulfilled Promise of Management as a Profession.* Princeton: Princeton University Press.

Kollar, R. J., & Green, S. L. (2008) 'Sarbanes–Oxley Act of 2002.' In R. W. Kolb (ed.) *Encyclopedia of Business Ethics and Society*. Thousand Oaks, CA: Sage Publications, Inc. Available at: http://search.ebscohost.com/login.aspx?direct=true&db=nlebk&AN=474262&site=ehost-live (Accessed 13 July 2019).

La Nouvelle Tribune (2018) 'Corruption – Numéro vert: 600 appels dénonciateurs reçus en quelques jours,' *La Nouvelle Tribune*. [online] Available at: https://lnt.ma/corruption-numero-vert-600-appels-denonciateurs-recus-quelques-jours/ (Accessed 15 July 2019).

LDIIfrane.org. (n.d.). [online] Available at: www.ldiifrane.org/about-us (Accessed 9 June 2019).

Le Desk (2018) 'El Otmani: le Maroc perd jusqu'à 7% du PIB à cause de la corruption,' *Le Desk*. [online] Available at: https://ledesk.ma/encontinu/el-otmani-le-maroc-perd-jusqua-7-du-pib-cause-de-la-corruption/ (Accessed 15 July 2019).

Leuz, C. (2007) 'Was the Sarbanes–Oxley Act of 2002 Really This Costly? A discussion of evidence from event returns and going-private decisions,' *Journal of Accounting and Economics*, 44(1), pp. 146–165. doi: 10.1016/j.jacceco.2007.06.001.

Lopez, Y. P., & Martin, W. F. (2018) 'University Mission Statements and Sustainability Performance,' *Business and Society Review*, (2), p. 341. doi: 10.1111/basr.12144.

Makinwa, A. (2013) *Private Remedies for Corruption: Towards an International Framework*. The Hague, The Netherlands: Eleven International Publishing. Available at: http://search.ebscohost.com/login.aspx?direct=true&db=nlebk&AN=884437&site=eds-live (Accessed 8 November 2019).

MAPNews (2018) 'HM the King: Fighting Corruption, a Scourge that Hinders Economic and Social Advancement, Should be Made a Priority,' *MapNews*. [online] Mapnews.ma. Available at: www.mapnews.ma/en/activites-royales/hm-king-fighting-corruption-scourge-hinders-economic-and-social-advancement-should (Accessed 15 July 2019).

Marwell, G., & Ames, R. E. (1981) 'Economists Free Ride, Does Anyone Else?,' *Journal of Public Economics*, 15, pp. 295–310.

Mason, R. O. (2018) 'Scandals, Corporate.' In R. W. Kolb (ed.) *Encyclopedia of Business Ethics and Society*. Thousand Oaks, CA: Sage Publications, Inc. Available at: http://search.ebscohost.com/login.aspx?direct=true&db=nlebk&AN=474262&site=ehost-live (Accessed 13 July 2019).

Medias24 (2016) Numéro vert anti-corruption: 12 condamnations sur 200.000 appels. [online] Available at: www.medias24.com/MAROC/DROIT/168239-Le-numero-vert-anti-corruption-est-un-echec-12-condamnations-sur-200.000-appels.html (Accessed 14 July 2019).

Melé, D. (2008) 'Integrating Ethics into Management,' *Journal of Business Ethics*, 78(3), p. 291. https://doi.org/10.1007/s10551-006-9343-7.

Mmsp.gov.ma. (2019) [online] Available at: www.mmsp.gov.ma/uploads/file/Rapport_Cnac_1erePhase_SNLCC_VD.pdf?fbclid=IwAR38En7XVUq-NpGbOS4vdKuLNTewz4OOP06bYZp3iKuWD69GNVhSx55NcoE (Accessed 15 July 2019).

Moberg, D. J. (2006) 'Best Intentions, Worst Results: Grounding Ethics Students in the Realities of Organizational Context,' *Academy of Management Learning and Education*, 5(3), pp. 307–316.

Oddo, A. R. (1997) 'A Framework for Teaching Business Ethics,' *Journal of Business Ethics*, 16(3), 293.

Onumah, J. M., Antwi-Gyamfi, N. Y., Djin, M., & Adomako, D. (2012) 'Ethics and Accounting Education in a Developing Country: Exploratory Evidence from the

Premier University in Ghana.' [online] Ugspace.ug.edu.gh. Available at: http://ugspace.ug.edu.gh/handle/123456789/6478 (Accessed 13 July 2019).

Park, H. J. (1998) 'Can Business Ethics Be Taught? A New Model of Business Ethics Education,' *Journal of Business Ethics*, 17(9/10), p. 965.

PRME – Principles for Responsible Management Education (2019) [online] Unprme.org. Available at: www.unprme.org/ (Accessed 15 July 2019).

Ribstein, L. E. (2002) 'Market vs. Regulatory Responses to Corporate Fraud: A Critique of the Sarbanes–Oxley Act of 2002,' *Journal of Corporation Law*. http://search.ebscohost.com/login.aspx?direct=true&db=edsbl&AN=RN128021450&site=eds-live.

Roberts, C., & Roper, C. D. (2015) 'Ethics for Students Means Knowing and Experiencing: Multiple Theories, Multiple Frameworks, Multiple Methods in Multiple Courses.' In D. E. Palmer (ed.) *Handbook of Research on Business Ethics and Corporate Responsibilities* (pp. 153–178). Hershey, PA: IGI Global.

Ronald, F. D., & Ragatz, J. A. (2008) 'Business, Purpose of.' In R. W. Kolb (ed.) *Encyclopedia of Business Ethics and Society*. Thousand Oaks, CA: Sage Publications, Inc. Available at: http://search.ebscohost.com/login.aspx?direct=true&db=nlebk&AN=474262&site=ehost-live (Accessed 13 July 2019).

Sauser, W. I., & Sims, R. R. (2015) 'Techniques for Preparing Business Students to Contribute to Ethical Organizational Cultures.' In D. E. Palmer (ed.) *Handbook of Research on Business Ethics and Corporate Responsibilities* (pp. 221–248). Hershey, PA: IGI Global.

Schoenfeldt, L. F., McDonald, D. M., & Youngblood, S. A. (1991) 'The Teaching of Business Ethics: A Survey of AACSB Member Schools,' *Journal of Business Ethics*, 10(3), p. 237.

Schwartz. M. S. (2008) 'Business Ethics.' In R. W. Kolb (ed.) *Encyclopedia of Business Ethics and Society*. Thousand Oaks, CA: Sage Publications, Inc. Available at: http://search.ebscohost.com/login.aspx?direct=true&db=nlebk&AN=474262&site=ehost-live (Accessed 13 July 2019).

Simpson, S. N. Y, Onumah, J. M., & Oppon-Nkrumah, A. (2015) 'Ethics Education and Accounting Programmes in Ghana: Does University Ownership and Affiliation Status Matter?,' *International Journal of Cyber Ethics in Education*, 1(1), pp. 43–56.

Sims, R. R., & Felton, E. L. J. (2006) 'Designing and Delivering Business Ethics Teaching and Learning,' *Journal of Business Ethics*, 63(3), 297. https://doi.org/10.1007/s10551-005-3562-1.

Songwe, V. (2018) 30th African Union Summit, 32nd Ordinary Session of the Executive Council, Theme: Winning the fight against corruption: A sustainable path to Africa's transformation. [online] uneca.org. Available at: https://repository.uneca.org/bitstream/handle/10855/24240/b1188308x.pdf?sequence=1 (Accessed 6 June 2019).

Spector, B. I. (2012) *Detecting Corruption in Developing Countries: Identifying Causes/strategies for Action*. Sterling, VA: Kumarian Press. Available at: http://search.ebscohost.com/login.aspx?direct=true&db=nlebk&AN=440899&site=eds-live (Accessed 8 November 2019).

SSIR.org (2019) Giving Voice to Values (SSIR). [online] Available at: https://ssir.org/articles/entry/giving_voice_to_values (Accessed 15 July 2019).

Stachowicz-Stanusch, A., & Amann, W. (2019) *Anti-corruption in Management Research and Business School Classrooms*. Charlotte, NC: Information Age Publishing (Research in Management Education and Development). Available at: http://search.ebscohost.com/login.aspx?direct=true&db=edsebk&AN=1937668&site=eds-live (Accessed 8 November 2019).

Store.darden.virginia.edu (2019) 'Business Case Studies & Business Publications – Darden Business Publishing. Giving Voice to Values' [online] Available at: http://store.darden.virginia.edu/giving-voice-to-values (Accessed 13 June 2019).

Suyantiningsih, S. & Rahmadonna, S. (2019) 'Addie Model: Development of Anti-Corruption Education Materials in Elementary School,' *KnE Social Sciences & Humanities*, (July): 389. Available at: http://search.ebscohost.com/login.aspx?direct=true&db=edb&AN=138196225&site=eds-live.

Tavanti, M., & Wilp, E. A. (2015) 'Globally Responsible Management Education: From Principled Challenges to Practical Opportunities.' In D. E. Palmer (ed.) *Handbook of Research on Business Ethics and Corporate Responsibilities* (pp. 196–220). Hershey, PA: IGI Global.

Telquel.ma. (2018) "Numéro vert" contre la corruption: seulement 36 poursuites en deux ans et demi. [online] Available at: https://telquel.ma/2018/01/24/numero-vert-contre-corruption-36-poursuites-en-ans-demi_1577597/?utm_source=tq&utm_medium=normal_post (Accessed 14 July 2019).

thecasecentre.org (2019) 'The Case Centre homepage' [online] Available at: www.thecasecentre.org/main/ (Accessed 15 July 2019).

Tran, B. (2015) 'Bridging the Foundational Gap between Theory and Practice: The Paradigm on the Evolution of Business Ethics to Business Law.' In D. E. Palmer (ed.) *Handbook of Research on Business Ethics and Corporate Responsibilities* (pp. 123–152). Hershey, PA: IGI Global.

Transparency International (2016) 'People and Corruption: Middle East and North Africa Survey' [Online] Available at www.transparency.org/whatwedo/publication/people_and_corruption_mena_survey_2016.

Transparency International (2018) 'Corruption Perceptions Index 2018' [online] www.transparency.org. Available at: www.transparency.org/cpi2018 (Accessed 15 July 2019).

Transparency International (2019) 'Transparency International – The Global Anti-Corruption Coalition' [online] Transparency.org. Available at: www.transparency.org/ (Accessed 15 July 2019).

Transparencymaroc.ma (2019) 'Transparency Maroc' [online] Available at: http://transparencymaroc.ma/TM/ (Accessed 15 July 2019).

Ugrin, J. C, & Odom, M. D. (2010) 'Exploring Sarbanes–Oxley's Effect on Attitudes, Perceptions of Norms, and Intentions to Commit Financial Statement Fraud from a General Deterrence Perspective,' *Journal of Accounting and Public Policy*, 29(5), pp. 439–458. doi: 10.1016/j.jaccpubpol.2010.06.006.

Unglobalcompact.org (2019) 'UN Global Compact' [online] Available at: www.unglobalcompact.org/ (Accessed 15 July 2019).

Wolff-Mann, E. (2018) 'Facebook Employee Morale is Low' [online] Available at: https://finance.yahoo.com/news/facebook-employee-morale-low-133129872.html. (Accessed 26 March 2020).

World Economic Forum (2014) 'The Global Competitiveness Report 2013–2014' [online] Available at: www.weforum.org/reports/global-competitiveness-report-2013-2014 (Accessed 15 July 2019).

Zhang, J., & Han, J. (2016) 'Adoption of Sarbanes–Oxley Act in China: Antecedents and Consequences of Separate Auditing,' *International Journal of Auditing*, (2), p. 108. doi: 10.1111/ijau.12057.

6 Integrating sustainability and CSR concepts in the College of Business and Economics (CBE) curriculum

An experiential learning approach

Dalia Abdelrahman Farrag and Shatha Obeidat

Introduction

As a college that is accredited by the Association to Advance Collegiate Schools of Business (AACSB), the College of Business and Economics (CBE), in Doha, Qatar is required to provide students "tools for recognizing and responding to ethical issues, both personally and organizationally" (AACSB Ethics Education Task Force, 2004, p. 9). For three consecutive years, the CBE has been collaborating with the Qatar Green Building Council (QGBC), a member of Qatar Foundation, to integrate sustainability and CSR concepts in some of its marketing and management curriculums. The main objectives of this collaboration is to increase the awareness level of students about sustainability and how they can integrate it in their daily lives, as well as enabling them to participate in competitions related to proposing sustainable solutions and ideas to Qatari companies. Students were required to work in hands-on projects and case studies related to sustainability and CSR as well as present their ideas to their fellow colleagues, faculty, and university leadership, to the annual Qatar Green Building Conference, and to designated clients that they worked with. Students were also responsible for launching the "Green Life Campaign" at the CBE for three consecutive years, as part of Qatar Sustainability Week, which receives a statewide recognition.

This chapter focuses on a case analysis related to a real-world project that management and marketing students were required to work on for an entire semester (15 weeks). The client was Nakilat. Nakilat is a Qatari-owned shipping and maritime company providing the critical transportation link in the State of Qatar's liquefied natural gas (LNG) supply chain. Management students were required to analyze policies and procedures related to sustainability at Nakilat to propose improvement plans. Marketing students were required to design an integrated marketing communications (IMC) campaign for Nakilat employees to encourage them to adopt sustainable behavior and perform activities in their daily work routine related to sustainability and be more socially responsible. These projects were part of Qatar Sustainability Week, an annual initiative organized and supported by the Qatar Green Building Council (QGBC). The College of Business and Economics (CBE) has been taking part in this week for three consecutive years with different ideas and student involvement.

This sustainability initiative by the CBE is very much related to the Giving Voice to Values (GVV) approach that has been developed by Mary Gentile of the University of Virginia Darden School of Business www.GivingVoice ToValues.org). The GVV approach encourages individuals to learn how to engage in communication or action that reflects their values within an organization's value system. The purpose of this chapter is to demonstrate how the experiential learning method utilized empowers students to adopt sustainable, ethical, and socially responsible actions and recommend the same for the companies that they are working with as part of their course requirements. This has been done through analyzing to what level the CBE at Qatar University integrates ethics, corporate social responsibility, and sustainability (ECSRS) concepts in its curriculum by providing a case-based example (the Nakilat project). Furthermore, a preliminary study was conducted to examine students' perception regarding ECSRS integration in the curriculum and their perception towards the different learning methods adopted to teach ECSRS concepts.

The main contribution of this chapter focuses on providing students the appropriate learning environment and experience to support the implementation of the GVV concept. GVV is a skill-based approach (Warnell, 2011); students acting on their skills, knowledge, and understanding are important prerequisites.

The authors have adopted an approach similar to the three As approach proposed by Gentile et al. in 2015 to conceptualize ethics in business curriculum. In phase 1, awareness, students understand and define ethical and sustainability issues. In phase 2, analysis, students worked on a real case study (Nakilat Company) and proposed ideas and solutions. Finally, in phase 3, action, students presented their action plans for Nakilat employees and demonstrated their willingness to adopt sustainable behavior.

Literature review

Background of sustainability concept

There is an increasing emphasis today in education for sustainability. Sustainability refers to "meeting the needs of the present generation without compromising the ability of future generations to meet their own needs" (Brundtland et al., 1987, p. 16). The definition implies the requirement to promote sustainable practices by individuals, organizations, and societies in a way that will help protect the natural environment and other resources for future generations (Kemp et al., 2005). It encompasses three main pillars that include not only the ecological system but also the economic and social systems of the environment (Von Der Heidt & Lamberton, 2011). Hence, taking care of sustainable issues implies the need to make proper decisions that take into account the interdependent relationship among the ecological, social, and economic systems.

Business companies give emphasis to sustainable practices, like corporate social responsibility (CSR). CSR is defined as "the management of the company's positive impact on society and the environment through its operations, products, or services and through its interaction with key stakeholders, such as employees, customers, investors, and suppliers" (IISD, 2004, p. 111). Companies today see the long-term value of investing in socially responsible practices, particularly how profitability and other long-term goals could be thought of in the larger context of social responsibility (Hellsten & Mallin, 2006). Consequently, companies today drive their progress toward innovative sustainable solutions (Hanelt et al., 2017). They are required to respond to the needs of several stakeholders to maintain positive relationships with them. This is well supported by a large body of evidence in scholarly literature showing that different stakeholders (e.g., customers, employees, investors) reward companies that engage in CSR activity (Bhattacharya et al., 2009). For example, many companies today thrive to become an employer of choice, and increasingly people choose companies that demonstrate socially responsible practices (Hancock, 2004). Accordingly, business schools are asked to respond to sustainability requirements by finding ways to incorporate sustainability and CSR practices into their curricula.

Integrating sustainability into business education

Overall, sustainability has become a global business practice. However, companies have also been struggling to consider corporate sustainability at the strategic level (Kiron et al., 2012). Given its importance, and in order to cope with the challenges of proper integration, there has been increased scientific interest in integrating corporate sustainability in companies' strategy, culture, and many of their operational activities (Engert et al., 2015). Accordingly, it has become essential for educational institutions to revise their strategies, restructure their courses, and rebuild their programs to reflect teaching sustainability (Von Der Heidt & Lamberton, 2011; Barber et al., 2014). Specifically, to the extent this is happening, sustainability issues are becoming more established in business education. Business schools integrate sustainability issues in their curriculum in order to increase students' awareness of corporate social and ethical issues (Setó-Pamies & Papaoikonomou, 2016). While business ethics represents the moral reflection about corporate practices, social responsibility reflects the outcome of corporate actions entailing economic, ecological, and social consequences (Setó-Pamies et al., 2011). Both concepts are interrelated since corporate responsibility requires ethical corporate practices. The main goal is to educate future managers and employees on how to make ethical decisions. Furthermore, students need to develop their skills in business ethics and CSR in order to meet the challenges of the real business world (Setó-Pamies et al., 2011; Gottardello & Pàmies, 2019).

Business schools rely on different perspectives of integrating sustainability into their curricula. Rusinko (2010) proposed a matrix to integrate social responsibility in business education. This matrix identified four ways as an opportunity for integrating sustainability in management and business education. "Piggybacking" focuses on the integration of sustainability within existing structure by adding sustainability to individual courses or modules. Using this approach, a large number of students can learn about CSR and business ethics (Painter-Morland et al., 2016). Moreover, soft skills, such as teamwork, can be taught to students in congruence with this approach. The "digging deep" approach integrates sustainability through new stand-alone modules or courses (Mburayi & Wall, 2018). Business schools that use this approach can offer new modules, such as an optional business ethics course. It is considered an easy technique for implementing sustainability. Although it can be a useful method for providing in-depth knowledge of sustainability, this method has been criticized since these courses are unlikely to be selected by the students who need them the most (Painter-Morland et al., 2016), especially if they are offered as elective courses. The "mainstreaming" approach, on the other hand, integrates sustainability into common core requirements. This may encompass the content of sustainability-related tools such as student placements within social enterprises (Godemann & Michelsen, 2011). Finally, the "focusing" perspective aims to set up a new program for sustainability or introduce a new module in all programs. The focusing approach allows business schools to include an interdisciplinary perspective of sustainability across the whole curriculum (Roome, 2005). Overall, the matrix identified by Rusinko (2010) is helpful in understanding how sustainability can be integrated in business curricula.

Regardless of the method used to integrate sustainability into the business curriculum, the sustainability concept needs to be related to a specific business field, such as organizational behavior, human resource management, accounting, or financial reporting (Barber & Venkatachalam, 2013). Moreover, proper learning methods are required to teach social responsibility in business programs, such as case studies, guest speakers, field trips, and open discussion (Pratap Singh et al., 2011). While case studies are the most common among business professors for integrating sustainability issues in business, a challenge remains on the proper choice and implementation of a pedagogical tool that is relevant to existing course delivery requirements.

Giving Voice To Values (GVV) approach

GVV is a recent approach to business ethics education introduced by Gentile (2010b, 2010a). The main goal of the GVV pedagogical approach is to enable people to make positive change inside organizations by using "problem definition, creative problem solving, constructive engagement, persuasion, reasoning, personal example and leadership" (Gentile, 2010b). This approach focuses on ethical implementation rather than solely focusing on ethical analysis (Shawver & Miller, 2019). Particularly, traditional approaches to business ethics education focus on two things: building awareness and ethical analysis. It aims to build awareness by business students of ethical issues so that they can recognize them when they encounter them (Gentile et al., 2015). Moreover, it prepares them for consistent analysis of ethical situations by introducing students to ethical reasoning. The GVV approach goes further by adding a third element called "action" to the traditional emphasis on awareness and analysis, that is, the course of action and the implementation of ethical solutions (Gentile, 2017; Shawver & Miller, 2019). This important element shifts the way business students learn how to deal with ethical dilemmas (Gentile et al., 2015).

The adoption of the GVV approach is flexible and can allow for the use of existing learning methods for teaching ethics and sustainability issues (Levesque, 2018). For example, existing case studies can be used and modified to suit the GVV concept. In particular, an effective GVV case study needs to ask students, after analyzing a situation, to discuss "how" not "whether" to get the right thing done (Arce & Gentile, 2015). Other methods could be employed as part of the GVV learning, such as team projects and assignments asking students to prepare a script or action plan for implementation (Gentile, 2017). The focus on "action" can consequently lead to building students' confidence and competence to actually enact their values (Painter-Morland & Slegers, 2018).

The GVV approach fills in the action gap in teaching business ethics. Its level of application in business schools is well known, since it has been piloted in over 1,065 schools, companies, and other organizations on six continents (Giving Voice to Values – Darden School of Business, 2019). Moreover, some research has stressed the positive impact of the GVV approach (Christensen et al., 2016) and specifically in business education (Warnell, 2011). This research provides a case of one of the leading business schools in the MENA, particularly in the Gulf region.

College of Business and Economics in Qatar University: case study

The world's leading international body for accreditation for business schools, the Association to Advance Collegiate Schools of Business (AACSB), states that the curriculum content for the bachelor degree and higher should include business knowledge of social responsibility, including sustainability, diversity, and ethical behavior and approaches to management (AACSB, 2013). Hence, meeting the AACSB requirements urges business schools to place more emphasis on integrating sustainability into their curriculum. The

College of Business and Economics in QU is one of nine out of 450 Arab universities in the Middle East that have achieved the AACSB accreditation. In line with the requirements of its accredited status, the CBE continued its efforts of re-evaluating its programs and academic and research goals in order to meet the AACSB criteria. Accordingly, this requires that the CBE focus should be positioned on integrating social responsibility and ethics into its programs.

With its vision "To be the destination of excellence in education, scholarship, and practice in business and economics," the CBE strives to provide its students and faculty with optimum opportunities to achieve international standards of excellence (CBE, 2019). Based on 2018 records, the CBE employs more than 100 faculty members and has more than 4,673 undergraduate students and 171 graduate students enrolled in it (CBE, 2019). The college encompasses three departments: management and marketing, finance and economics, and accounting and management information systems, with the Department of Management and Marketing being the largest department in terms of the number of students.

In order to integrate sustainability and ethics in the college program, the college offers the BBA (Bachelor of Business Administration) program, which incorporates two sustainability-related objectives: appreciate social responsibilities of business decision-making and appreciate ethical dimensions of business decision-making. Moreover, these two learning outcomes are assessed in every academic cycle as part of the college assessment plan. As part of its effort to integrate sustainability and ethics in its curriculum, the CBE makes use of the "piggybacking" approach that focuses on the integration of sustainability within existing structure by adding ethics and sustainability issues to individual business and management courses and modules. Particularly, the majority of business courses include sustainability materials that are presented and discussed during the class. Moreover, instructors rely on other learning methods like case studies and guest speakers that discuss sustainability-related concepts.

To further investigate the kind of integration level of sustainability and ethics issues in its business curriculum, the current study identified two core objectives. Since we believe that student experience is vital and should be considered, the first one looks into whether current undergraduate students in the CBE are aware of the ECSRS concepts that could be integrated in the business curriculum, and on the type of learning methods they believe are relevant to be used for teaching sustainability issues in business. The second looks to provide a real case example of a teaching tool used in order to discuss the level of integration reached and possible venues for improvement to apply the more action-oriented approach, the GVV approach. Simply, our perspective could be summarized in the proposed framework illustrated in Figure 6.1, consisting of three different phases: awareness, analysis, and action.

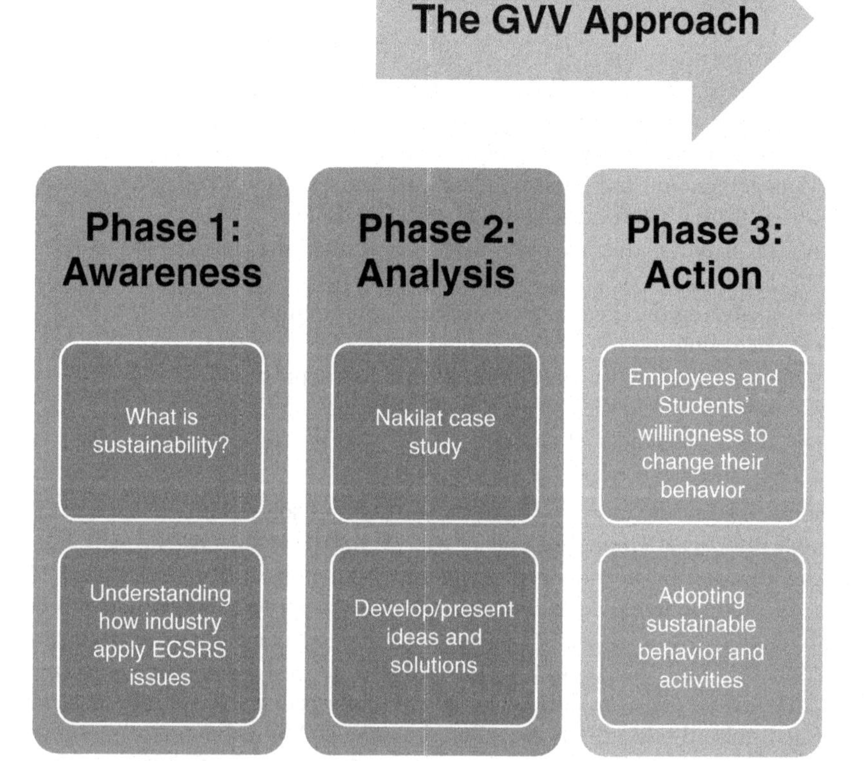

Figure 6.1 Case study framework.

Source: adapted from Gentile et al. (2015).

Methodology

Case study analysis

Based on the case study framework, the awareness phase has been fulfilled using a two-fold approach. First, the course outlines have been modified to include ECSRS concepts in the form of a term project for the promotion management and organizational behavior curriculums. The term project adopted the real case study approach and the client was Nakilat. Second, Nakilat public relations and human resources managers delivered different information and awareness sessions to students about how Nakilat implements ECSRS concepts, the different challenges that the company is facing in doing so, and the objectives from both the marketing and management perspectives. More specifically, this case study analysis covered the following objectives:

- Marketing-related objectives (promotion management course):
  - to raise awareness about sustainability in the workplace;
  - to change the employees' beliefs and attitude towards sustainability (to be positive);
  - engage employees into playing an active role for inspiring sustainability at the workplace;
  - continuously motivating employees to continue being sustainable.
- Management-related questions (organizational behavior course):
  - analyzing work ethic and social responsibility;
  - identifying the role of the leader and the leadership patterns followed;
  - analyzing the level of motivation amongst employees;
  - analyzing the role of developing teams;
  - understanding the role of decision-making process;
  - understanding the role of organizational design;
  - the management of change.

Students adopted different methods to analyze and solve the case study. Initially, all students attempted to collect data about the case study through conducting desk research and distributing surveys amongst Nakilat employees. Then, through applying course material, students proposed different ideas from the marketing and management perspectives to encourage Nakilat employees to adopt sustainable behavior. Students worked in groups and were organized as if they were a real consulting company. The flipped classroom approach has also been utilized using the peer-assisted learning (PAL) technique to reinforce course content and allow students to learn in a variety of different ways. Finally, students were required to present their ideas using an infographic poster at the "Green Corporate Campaign" that took place at the College of Business and Economics as part of Qatar Sustainability Week. A total of 365 students forming 32 groups from five classes participated in the real case study analysis. Assessment criteria included factors related to creativity, planning, strategy, tools utilized, and actual implementation of ideas on a five-point Likert scale.

Quantitative research

In order to assess the students' experience in the course after integrating sustainability concepts in the curriculum and applying what they have learned on Nakilat as a real case study, an empirical study in the form of a survey has been conducted. The empirical study sought to answer the following questions:

- How do business students perceive a variety of ethics, corporate social responsibility, and sustainability (ECSRS) topics that are integrated into the business curriculum?
- How do business students perceive the integration of ECSRS topics into the business curriculum?
- What are students' preference of teaching methods used to integrate ECSRS concepts into the business curriculum?
- What is the intention level of business students to adopt sustainable behaviors after being engaged in business courses that integrate ECSRS topics?

A convenience sampling technique has been adopted to ensure the eligibility and willingness of students to participate in the survey. The survey has been sent online to all students that participated in the case study analysis (i.e., a total of 365 students). A final sample of 222 usable questionnaires were gathered and used in the analysis.

The scales used for measuring perception of students towards relevance of sustainability topics to the business curriculum and methods of teaching used has been adopted from Barber and Venkatachalam (2013). Students' perceptions about integrating ECSRS topics in the business curriculum have been measured by a scale adopted from Albaum and Peterson (2006). Finally, intentions to adopt sustainable behavior after the course experience has been measured using a scale adopted from Swaim et al. (2014).

Findings

The findings of the empirical study were based on a statistical analysis using SPSS software. Multiple checks for validity and reliability were performed. In particular, Cronbach's alpha values were calculated for reliability measures (Hair et al., 2009). All Cronbach's values for the study constructs are above the minimum acceptable threshold of 0.70. The results showed that the study constructs meet the reliability tests. Moreover, in order to evaluate content validity, care was taken to provide careful definition questions and items used in the survey and all items were obtained from previously used scale measures. The questionnaire was pilot-tested by two academics and students before use and was modified accordingly.

Demographic characteristics are shown in Table 6.1. The majority of students are female (78.8%) and are in the age group of 22–25 years old (41.4%), which is very much representative of the student structure at the College of Business and Economics (CBE). Female students at the CBE currently constitute more than 70% of the total number of students (4,673), which justifies why the study sample is skewed more towards females. The different majors have also been represented at the CBE: Management (64%), Marketing (22.5%), Accounting (5%), Finance (4.5%), Economics (2.3%), and MIS (1%), as can be seen in Table 6.2.

Table 6.1 Demographic characteristics

		Frequency	*%*
Valid	Male	47	21.2
	Female	175	78.8
Total		222	100.0
Valid	18–21	87	38.3
	22–25	92	41.4
	26 or older	43	19.4
Total		222	100.0

Table 6.2 Distribution of participated students by study major

Major		*Frequency*	*%*
Valid	Management	142	64.0
	Economics	5	2.3
	Marketing	50	22.4
	Management Information Systems	4	2
	Finance	10	4.5
	Accounting	11	5.0
Total		222	100.0

Phase 1: Awareness

Students have demonstrated their findings for the awareness phase through deep understanding of what sustainability is and how the client Nakilat applies ECSRS standards at different levels and departments. They also relied on the different kinds of research conducted for understanding the behavior of employees towards ECSRS concepts. Furthermore, awareness has been measured through perceptions of students towards the different ECSRS concepts that could be integrated in business curriculum. Table 6.3 highlights the main findings.

Table 6.3 Students' perception of sustainability topics relevance

Subjects	*Mean*	*Std. Deviation*
• Green Innovation (Technology to Market)	4.06	0.864
• Green Global Market Development	4.01	1.002
• Green Entrepreneur	4.05	0.994
• Product Prototype Design	3.94	0.987
• Green Information Technology Solutions	3.86	1.030
• Environmental Auditing	3.91	1.128
• Sustainable Business Reporting	3.95	0.964
• Sustainable Development Concepts	4.27	0.862

Subjects	*Mean*	*Std. Deviation*
• Natural Resource Economics	4.23	0.944
• Personal Ethics and Values	4.37	0.946
• Consumer Behavior and Attitudes	4.28	0.982
• Environmental Policy Development	4.04	1.015
• Socially Responsible Investing	4.14	0.937
$n = 222$		

As part of the CBE curriculum in Qatar University, our preliminary study findings demonstrated that undergraduate business students perceive the following as the most relevant sustainability topics to be offered by the CBE: personal ethics and values (mean=4.37), consumer behavior and attitudes (mean=4.28), sustainability development concepts (mean=4.27), natural resource economics (mean=4.23), and socially responsible investing (mean=4.14). Other topics were also found to be relevant, such as green innovation, green entrepreneur, environmental policy development, and green global market development. The least relevant topic was viewed to be "green information technology solutions." The results can be seen in Table 6.3.

It is therefore reasonable to conclude that business sustainability has a high profile amongst QU undergraduate business students, which has increased after this case study analysis. Moreover, sustainability topics that are viewed by students to be valid cover a variety of business fields (i.e., management, marketing, finance, etc.). Accordingly, it can be concluded that business students are aware of the importance of these sustainability topics as part of the business curriculum.

Phase 2: Analysis

Marketing students developed creative campaign ideas for motivating and encouraging Nakilat employees to adopt sustainable behavior and perform sustainable activities at work and in their daily lives. The three winners were the "Charge it," "Move it," and "Click" campaigns. The three campaigns were built on simple and creative ideas that proposed solutions for the case study. Students used an A1 infographic to summarize their campaign ideas from awareness to behavior.

On the other hand, management students examined possible ways to improve how employees' behaviors are managed at Nakilat. They studied different aspects at work such as leadership, teamwork, motivation, communication process, individual differences, organizational design, etc. Based on their analyses, they recommended solutions on how to improve the management process and possible action plans for implementation. Management students displayed their work to a number of academic and professional audiences. Based on the jury evaluation and the previously mentioned assessment criteria, three teams won in the competition.

Furthermore, in order to assess the students' experience in the courses, they were asked to answer questions using a survey related to the teaching methods preferred for understanding ECSRS concepts as well as their perceptions towards integrating these concepts in the business curriculum.

Table 6.4 highlights the main findings for the most preferred teaching method for learning about ECSRS concepts as perceived by students. For this purpose, the scale adopted used a five-point scale, where 1 means that this method is least preferred and 5 means that is most preferred, and numbers in between resemble different levels of preference. The results showed that students mostly prefer short-term visits to companies that adopt sustainability (mean = 4.18), followed by games and simulation techniques (mean = 4.15) as a learning method for teaching sustainability, then direct interaction with managers dealing with sustainability (mean = 4.10). The least preferred learning method for students was presentations to students (mean = 3.61). Other unconventional methods like multimedia presentations, real case studies, and group projects also showed high preference amongst students as compared with traditional methods of teaching.

Finally, Table 6.5 presents findings related to students' perception of integrating ethics, social responsibility, and sustainability concepts in the CBE curriculum. For this purpose, the scale adopted used a five-point Likert scale, where 1 means strongly disagree and 5 means strongly agree. In general, CBE students agree to integrate ECSRS concepts in the business curriculum

Table 6.4 Preferred learning methods for sustainability

Learning Method	*Mean*	*Std. Deviation*
• Clarify the concept of sustainability through multimedia presentations	3.96	0.806
• Incorporate sustainability issues through case studies	3.94	0.862
• Interacting with managers dealing with sustainability in real environments	4.10	0.801
• Short-term visits for students to organizations that are actively sustainable	4.18	0.888
• Student exchange programs for sharing ideas on sustainability	3.77	1.071
• Guest lectures from different professors and managers on sustainability	3.87	0.954
• Presentations to students on selected issues around sustainability	3.61	1.061
• Group projects on sustainability with a deliverable to a business/organization	3.97	0.823
• Games and/or simulation techniques to clarify sustainable-related concepts	4.15	0.831

$n = 222$

Table 6.5 Integrating ethics and CSR concepts in CBE curriculum

Statement	*Mean*	*Std. Deviation*
• My business education is preparing me to manage business and societal issues.	4.00	0.752
• My business school has ethics and sustainability aspects integrated in general curricula.	3.86	0.803
• Sustainability is something which all business courses should actively incorporate and promote.	4.26	0.821
• Sustainability is something which I would like to learn about.	4.19	0.858
• Business school should incorporate sustainability into its curricula in order to enhance student employability.	4.10	0.877
• Linking sustainability issues in educational curricula is interesting and relevant.	4.11	0.806
• We need to be aware of sustainability issues.	4.35	0.680
• We need to be thoroughly informed about CSR.	4.27	0.771
• Integrating CSR is essential for business curricula.	4.07	0.824
• Employers will look for knowledge on sustainability in my educational background.	3.61	0.929
Average	4.08	
$n=222$		

(overall mean of 4.08). Particularly, students emphasize the need to be thoroughly aware of CSR (mean = 4.35) and that they should be informed about CSR (mean = 4.27). Also, they believe that sustainability is something all business courses should incorporate (mean = 4.26). Finally, students moderately agree that that business school integrates sustainability aspects in the curricula (mean = 3.86), indicating that more needs to be done regarding this aspect across different curriculums.

Phase 3: Action

The final stage of the conceptual framework proposed, which is the action phase, has been supported as well. This is the phase where the GVV approach shines as students start transforming their plans into action through creatively presenting and demonstrating their ideas to the client as well as getting involved in different sustainable endeavors. It is clear from the survey results that most of the students showed their willingness to adopt sustainable behavior and get involved in different sustainable activities, as highlighted in Table 6.6.

The results showed that students are planning to increase their sustainable activities (mean = 4.14) and to look at how they can play a role in protecting the environment (mean = 4.06).

Table 6.6 Intention to adopt sustainable behavior

Statement	*Mean*	*Std. Deviation*
• I plan to increase my sustainable activities (e.g., energy conservation, recycling) in the future.	4.14	0.865
• I intend to seek out more opportunities to be more sustainable in the future.	4.04	0.770
• In the future, I plan to look into how I can play a greater role in protecting the environment.	4.06	0.791
$n = 220$		

Implications and conclusion

Higher education in general can serve as a powerful tool to help create a more sustainable future. The approach adopted by the authors for integrating ECSRS concepts in business curriculum is very much similar to the "main-streaming" approach proposed by Rusinko (2010), which is related to adding sustainability concepts/learnings to the current course structure. Based on previous literature, the case study method is the most common approach for teaching sustainability issues; however, this chapter has added to this common approach a more "experiential learning" approach for students where they are working on a real case study for an existing and reachable company in the Qatari market: Nakilat.

The authors adopted the three As conceptual framework for solving the real case study, from awareness to analysis and finally action, as emphasized by the GVV approach (Gentile, 2017). Traditionally, most business ethics curriculums focused on phase one and two. However, based on the findings presented and the teaching methods and procedures adopted by the authors, all phases have been well executed and fulfilled. Students conducted research from various sources using different methods and had the opportunity to interact with the company employees and management throughout the semester and live the company experience to understand how Nakilat integrates ECSRS concepts in its policies, procedures, and corporate culture. Accordingly, students proposed creative and detailed campaigns and ideas that further stimulate the sustainable corporate culture at Nakilat and encourage employees to adopt sustainable behavior at work and in their daily lives. A few studies have stressed the importance of introducing learners to real-world examples as well as engagement with the industry for a better understanding of ECSRS issues (Robinson-Pant, 2009; Pratap et al., 2011; Warnell, 2011).

The winning teams also had the opportunity to get an internship at Nakilat and witness their ideas being put into action. Accordingly, an important implication is that the current learning methods adopted, generally speaking, should be improved to include an action plan dimension (or a script) for adoption of possible solutions and how students and employees can implement recommendations. There may also be other reasons/obstacles preventing the implementation of sustainability initiatives on campuses around the world. For example, related to the availability of inadequate conditions or favorable economic activities that will facilitate sustainability implementation (Pratap et al., 2011). However, this is not the case in our situation, as it is evident that Qatar University genuinely integrates sustainability initiatives in its vision and policies, from providing recycling bins across all colleges to encouraging faculty to conduct research about sustainability through the Center for Sustainability Development, which was established in 2014 at the university. Unfortunately, most of the research conducted is focusing on the engineering and petroleum industry. Furthermore, most university initiatives failed to go the extra mile of empowering students to voice and act on their values effectively.

In general, the use of the GVV approach in teaching can build greater openness and self-awareness amongst students to ethical considerations, as is evident from this case study analysis. However, can the GVV approach be implemented using any teaching method? Based on the findings of the case study analysis and the post-course survey distributed amongst students, the experiential learning method is the most preferred for teaching ECSRS concepts as compared with conventional methods like conducting presentations and inviting guest lecturers to talk about sustainability. Experience-based learning is also an important component of the GVV concept (Gentile 2010a, pp. 228–229). Findings emphasized the importance of the learning method for understanding, analyzing, and implementing ECSRS concepts. Furthermore, the students showed their willingness to adopt sustainable behavior and perform activities to act on their values after living the course experience.

The authors recommend that the College of Business and Economics improves their assessment methods in order to better match the two sustainable-related objectives (appreciate social responsibilities and appreciate ethical dimensions of business decision-making). Curriculum is primarily the responsibility of the faculty, thus they must initiate change. Business faculty should develop and create a sustainable learning culture on campus amongst students. Rewards and peer appreciation could also be supportive ideas for encouraging students and employees to take their ECSRS values from awareness to action. Business educators should consider delivering a Giving Voice to Values (GVV) elective course to undergraduate students and utilizing unconventional teaching methods to deliver it. Students and youth in general are questioning everything around them, and it is important to provide reasoning and rationales that allow them to explore and act on their values as emerging leaders. The GVV concept shall allow them to bridge the gap between intention and action; however, and as evident from this case analysis, the teaching method adopted plays and integral role in the success of the GVV pedagogy.

References

AACSB. 2013. 2013 Eligibility Procedures and Accreditation Standards for Business Accreditation Available: www.aacsb.edu/-/media/aacsb/docs/accreditation/business/standards-and-tables/2018-business-standards.ashx?la=en&hash=B9AF18F3FA0DF19B352B605CBCE17959E32445D9 (Accessed 2019).

AACSB Ethics Education Task Force Report. 2004. Available: https://www.aacsb.edu/~/media/AACSB/Publications/research-reports/ethics-education.ashx. (Accessed 26 March 2020).

Albaum, G., & Peterson, R. 2006. Ethical attitudes of future business leaders: Do they vary by gender and religiosity? *Business and Society*, 45(3), 300–321.

Arce, D. G., & Gentile, M. C. 2015. Giving voice to values as a leverage point in business ethics education. *Journal of Business Ethics*, 131, 535–542.

Barber, N. A., & Venkatachalam, V. 2013. Integrating social responsibility into business school undergraduate education: A student perspective. *American Journal of Business Education*, 6, 385–396.

Barber, A., Wilson, N., Venkatachalam, V., Cleaves, S., & Garnham, J. 2014. Integrating sustainability into business curricula: University of New Hampshire case study. *International Journal of Sustainability in Higher Education*, 15, 473–493.

Bhattacharya, C. B., Korschun, D., & Sen, S. 2009. Strengthening stakeholder–company relationships through mutually beneficial corporate social responsibility initiatives. *Journal of Business Ethics*, 85, 257–272.

Brundtland, G. H., Khalid, M., Agnelli, S., & Al-Athel, S. 1987. *Our Common Future.* New York: United Nations.

CBE. 2019. CBE Facts and Figures 2017–2018. Available at: www.qu.edu.qa/static_file/qu/colleges/cbe/documents/Public_Disclosure_of_CBE_Performance_AY_2017_2018.pdf.

Christensen, A., Cote, J., & Latham, C. K. 2016. Developing ethical confidence: the impact of action-oriented ethics instruction in an accounting curriculum. *Journal of Business Ethics*, 153, 1157–1175.

Engert, S., Rauter, R., & Baumgartner, R. 2015. Exploring the integration of corporate sustainability into strategic management: A literature review. *Journal of Cleaner Production*, 112.

Gentile, M. 2010a. Giving Voice To Values [Online]. Available at: www.givingvoicetovaluesthebook.com/ (Accessed 2019).

Gentile, M. C. 2010b. *Giving Voice to Values: How to speak your mind when you know what's right.* New Haven, CT: Yale University Press.

Gentile, M. C. 2017. Giving voice to values: a pedagogy for behavioral ethics. *Journal of Management Education*, 41, 469–479.

Gentile, M., Lawrence, A., & Melnyk, J. 2015. What is a giving voice to values case? *Case Research Journal*, 35, 1–10.

Giving Voice to Values – Darden School of Business. 2019. [Online] https://www.darden.virginia.edu/ibis/initiatives/gvv (Accessed April 2020).

Godemann, J., & Michelsen, G. 2011. Sustainability communication: an introduction. In *Sustainability Communication*. Berlin: Springer.

Gottardello, D., & Pàmies, M. D. M. 2019. Business school professors' perception of ethics in education in Europe. *Sustainability*, 11(3), 608.

Hair, J., Black, W., Babin, B., & Anderson, R. 2009. *Multivariate Data Analysis.* New York: Pearson Education.

Hancock, J. 2004. *Investing in Corporate Social Responsibility: A guide to best practice, business planning and the UK's leading companies*. London: Kogan Page.

Hanelt, A., Busse, S., & Kolbe, L. M. 2017. Driving business transformation toward sustainability: exploring the impact of supporting IS on the performance contribution of eco-innovations. *Information Systems Journal*, 27, 463–502.

Hellsten, S., & Mallin, C. 2006. Are 'ethical' or 'socially responsible' investments socially responsible? *Journal of Business Ethics*, 66, 393–406.

IISD. 2004. Perceptions and Definitions of Social Responsibility. Available at: http://inni.pacinst.org/inni/corporate_social_responsibility/standards_definitions.pdf.

Kemp, R., Parto, S., & Gibson, R. B. 2005. Governance for sustainable development: moving from theory to practice. *International Journal of Sustainable Development*, 8, 12–30.

Kiron, D., Kruschwitz, N., Haanaes, K., & Velken, I. 2012. Sustainability nears a tipping point. *MIT Sloan Management Review*, 53, 69–74.

Levesque, L. L. 2018. Student-authored ethics vignettes: Giving Voice to Values all semester. *Management Teaching Review*, 3, 331–345.

Mburayi, L., & Wall, T. 2018. *Sustainability in the Professional Accounting and Finance Curriculum: An exploration*. Available online at: https://chesterrep.openrepository.com/handle/10034/621192. (Accessed 26 March 2020).

Painter-Morland, M., Sabet, E., Molthan-Hill, P., Goworek, H., & De Leeuw, S. 2016. Beyond the curriculum: integrating sustainability into business schools. *Journal of Business Ethics*, 139, 737–754.

Painter-Morland, M., & Slegers, R. 2018. Strengthening "Giving Voice to Values" in business schools by reconsidering the "invisible hand" metaphor. *Journal of Business Ethics*, 147, 807–819.

Pratap Singh, T., Bisht, N., & Rastogi, M. 2011. Towards the integration of sustainability in the business curriculum: perspectives from Indian educators. *Journal of Global Responsibility*, 2, 239–252.

Robinson-Pant, A. 2009. Changing academies: exploring international PhD students' perspectives on 'host' and 'home' universities. *Higher Education Research and Development*, 28 (4), 417–429.

Roome, N. 2005. Teaching sustainability in a global MBA: insights from the OneMBA. *Business Strategy and the Environment*, 14, 160–171.

Rusinko, C. A. 2010. Integrating sustainability in management and business education: a matrix approach. *Academy of Management Learning & Education*, 9, 507–519.

Setó-Pamies, D., Domingo-Vernis, M., & Rabassa-Figueras, N. 2011. Corporate social responsibility in management education: current status in Spanish universities. *Journal of Management & Organization*, 17, 604–620.

Setó-Pamies, D., & Papaoikonomou, E. 2016. A multi-level perspective for the integration of ethics, corporate social responsibility and sustainability (ECSRS) in management education. *Journal of Business Ethics*, 136, 523–538.

Shawver, T. J., & Miller, W. F. 2019. *Giving Voice to Values in Accounting*. New York: Routledge.

Swaim, J. A., Maloni, M. J., Napshin, S. A., & Henley, A. B. 2014. Influences on student intention and behavior toward environmental sustainability. *Journal of Business Ethics*, 124, 465–484.

Von Der Heidt, T., & Lamberton, G. 2011. Sustainability in the undergraduate and postgraduate business curriculum of a regional university: A critical perspective. *Journal of Management & Organization*, 17, 670–690.

Warnell, J. M. 2011. Ask more of business education: Giving Voice to Values for emerging leaders. *Journal of Business Ethics Education*, 8(1), 320–325.

Conclusion

Corporate social responsibility (CSR) is a notion that is gaining momentum in different academic and practitioner circles. CSR (and CSR education) questions the capitalist ideology that has an "inbuilt imperative of (economic) growth" (Attfield, 2015: 85). Despite the complexity of reaching consensus regarding a formal definition for CSR, Sheehy (2015: 639) has reviewed prior definitions and defined it as "a socio-political movement which generates private self-regulatory initiatives, incorporating public and private international law norms seeking to ameliorate and mitigate the social harms of and to promote public good by industrial organizations." New management philosophies and paradigms have been developed to challenge the dominant social (capitalist) paradigm, such as the humanistic management philosophy (Dierksmeier, 2016). Windsor (2016: 135) notes that a global "moral science" for business ethics can stem from "the no harm axiom."

Still, the ability of business schools to tackle ethics and sustainability issues in CSR education has been questioned. Rozuel (2016), however, notes that imagination and creative material can have a great impact on developing and boosting ethical reflection. One such creative approach is, in fact, the Giving Voice to Values (GVV) approach. GVV is a pedagogy and philosophy that fosters values-driven leadership through creative case writing that is used by business (and non-business) schools to promote ethical reflection. Rather than presenting traditional ethical dilemmas, GVV starts on the premise that individuals know what is right and strive to achieve it, which aims at enhancing the well-being of individuals facing situations in which their values are challenged and questioned. Well-being is, in fact, a "business concept" and private businesses have a strong role to play in bolstering well-being (Durand & Boarini, 2016).

This book presents a portfolio of chapters from different prominent authors in the Middle East and North Africa (MENA) region that portray the relevance of CSR education to enhancing business ethics and the sustainability mindset in the region. The book starts with a citation-based systematic literature review of ethics, corporate social responsibility, and sustainability (ECSRS) education in the region as a general overview on the subject matter. Based on citation analysis, Chapter 1 presents leading journals, articles, and authors in the ECSRS field. The chapter also shows the state of development of ECSRS education in this novel context that was previously unexplored. The authors' research shows that the majority of publications in the region were conducted in the past five years with student perception studies being the most represented, yet the views of other stakeholders in the CSR ecosystem are under-represented. The research also recommends more ECSRS research output in unexplored countries in the region, including Algeria, Bahrain, and Kuwait, among others.

Chapter 2 focused on reviewing the literature on CSR in the developing countries as a nascent and growing field of study, out of which the MENA region studies stem. The chapter focuses on highlighting the state of CSR education in Egypt, as an important developing country in the region that is in transition. Based on the findings of the empirical study conducted, the chapter concluded with drawing implications for CSR education in the region, noting that higher education for sustainable development (ESD) is an important tool in raising awareness about the complexities of ECSRS issues.

Chapter 3 portrayed the actual implementation of the GVV approach in a business school in the MENA region to showcase the practicality of applying this creative educational tool. The chapter highlights that ethical scandals have reinforced the importance of embedding ethics education in business curricula. Using the GVV approach as a teaching philosophy, the authors present an outline for a business ethics course which can be relevant to all universities in this under-represented context.

Education for sustainable development (ESD) is then highlighted as a prominent notion in ECSRS education in the MENA region in Chapter 4. The chapter tackles the perceptions and understandings of sustainable development of individuals who have experienced ESD. Based on the study's results, the chapter highlights the profound role that ESD education can play in developing individual perceptions of the importance of incorporating sustainability values in their daily lives. The chapter concludes with showing the importance of community engagement, community-based learning, and university outreach for maximum societal impact, and for aiding universities, especially business schools, in becoming active and good citizens of society.

The last two chapters look at the role ESCRS education can play in influencing business students and the future business environment in the contexts of Morocco and Qatar. Chapter 5 highlights how ECSRS education, specifically business ethics education, can lead to future responsible managers and leaders, hence promoting more sustainable business environments. The chapter indicates that business ethics education is still at an early stage of development, with generally private schools more inclined to cater to the topic than public schools, especially those that are signatory to institutions and initiatives such as the United Nations – Principles of Responsible Management Education (UN-PRME). The chapter concludes by emphasizing that business integrity and the fight against business corruption can only be enhanced if business ethics education is promoted. The final chapter,

Chapter 6, adopts a new approach to ECSRS learning, namely experience-based learning, and performs a case study analysis to emphasize how novel approaches used in sustainability education can enhance the willingness of business students, as future leaders, to adopt sustainable behaviors.

The state of ECSRS education in universities can be relevant to many other countries in the MENA region. Higher education institutions (HEIs) in the region need to cascade down the topic of CSR, sustainability, and business ethics from their mission statements to their extracurricular activities. Policymakers, especially ministries for higher education in the designated countries of the region, need to enforce a stand-alone course on ECSRS in business schools for maximum impact. Business faculty members also have a strong role to play as they are the change agents that can bring about institutional change and carry forward the quest for a more moral and just world.

References

Attfield, R. (2015). Sustainability and Management, *Philosophy of Management*, vol. 14, pp. 85–93.

Dierksmeier, C. (2016). What is 'Humanistic' about Humanistic Management?, *Humanistic Management Journal*, vol. 1, pp. 9–32.

Durand, M., & Boarini, R. (2016). Wellbeing as a Business Concept, *Humanistic Management Journal*, vol. 1, pp. 127–137.

Rozuel, C. (2016). Challenging the 'Million Zeros': The Importance of Imagination for Business Ethics Education, *Journal of Business Ethics*, vol. 138, pp. 39–51.

Sheehy, B. (2015). Defining CSR: Problems and Solutions, *Journal of Business Ethics*, vol. 131, pp. 625–648.

Windsor, D. (2016). Economic Rationality and a Moral Science of Business Ethics, *Philosophy of Management*, vol. 15, pp. 135–149.

Index

Page numbers in **bold** denote tables, those in *italics* denote figures.

accredited/accreditation of business schools 52; AACSB, international body for 131–2; adopt stand-alone courses in business ethics 53, 114; agencies worldwide 93; College of Business and Economics (CBE), Qatar 127; requirement of business ethics courses in 114; status of CBE 132
accrediting agencies/bodies: of Al Akhawayn University (AUI) 110; EQUIS standards 107; in higher education 85, 93; highlight importance of mission statement 107; revised standards 93
Adler, P.S. 8, 97
Africa: developing country of Ghana in 94; Sub-Saharan, CSR reporting in 3; transformation of 99
African continent 3; Union summit 99
Africa, North/African, North *see* Middle East and North African (MENA) region
Aguilera, R. 9, 11, 21–2
Aguinis, H. 9–11, 21, 25
Ain Shams University, Egypt 58
Al-Abdin, A. 1, 14, 24–5, 56
Al Akhawayn University (AUI), Morocco 94, 110–12; Center for Business Ethics (CBE) 105, 113, 115, 120; Office of Community Involvement 111–12; Social Responsibility 111
anti-corruption: CGEM principles **103**; policies 100; principles of UN Global Compact **97**; strategy of Morocco *101*, 102–3
Arce, D.G. 1, 22, 62–4, 104–6, 131
Association for the Advancement of Sustainability in Higher Education (AASHE) 58–9
Association to Advance Collegiate Schools of Business (AACSB) 52, 93, 107, 127, 131; accreditation 132; Ethics Education Task Force 127

Bruntdland Report 71
business environment 146; code of ethics relevant to **61**; global, understanding **117**; institutional and stakeholder influences surrounding 44; promoting 98
business environment, clean: growing demand for 114; and uncorrupt, efforts to foster 100, 103
business environment, Egyptian **61**; challenges and obstacles embedded in 39
business environment in Morocco: corruption in 99; ethical landscape in 94; future 146; understanding **116**
business ethics education 10, 51–2; in accounting programs in Ghana 94; action-based approach to 103–4; challenges, attempt to mitigate 61; challenges and potential obstacles to 55; early stage of development 146; embedded in higher education system 57; faculty and student perspectives on 114; flexible approach in administration of 52–3; followed cognitive approaches and methods 53; GVV approach 130; importance in fight against corruption 114, 146; leading to future responsible managers

and leaders 146; mainstreaming 94, 109; Middle East 56; model for how business school can embrace 110; in multidimensional concept 115; rise of interest in 103; traditional approaches to 104, 131
business ethics education in Morocco 103, 106, 114; higher education institutions 94; status in 100
Business School Impact System (BSIS) 93
business schools 3; ability to tackle ethics and sustainability issues in CSR education 145; accredited 53; adopt principles of sustainability and social responsibility 92; asked to respond to sustainability requirements 129; contributed to developing PRME principles 98; corporate scandals casting shadow over probity of 96–7; criticized for promoting profit maximization 92; CSR curricula 58–9; curriculum, inclusion of ethics education advocated by AACSB 52; enforcement of stand-alone course on ECSRS 147; graduates 9, 20; GVV application in 131; importance of community engagement for aiding 146; integrate sustainability issues in their curriculum 129–30; interdisciplinary perspective of sustainability 130; introducing social issues courses 95; listed by Financial Times 9; in MENA region 131; pressured to address their educational shortcomings 97; programs, ethics education in 53; rating organization for accreditation of 52; role in promotion of ethical business conduct 103; stand-alone course on business ethics 114; strategic directions for 115; Western, emphasis on ethics 52; *see also* European business schools, Moroccan business schools, United States (US) business schools

Campbell, J. 35–8, 42
Center for Business Ethics (CBE), Al Akhawayn University: GVV curriculum translated into Arabic 105; new business ethics approach (GVV) 113
Central Authority for the Prevention of Corruption (*Instance Centrale de Prévention de la Corruption* (ICPC)) 100
Character Education Camp (CEC) **83**, 84
codes 80; of conduct 96; dress, for employees 57
code (s) of ethics **60**, 115; business 104; corporate **61**; developing for organizations 61; development 60
College of Business and Economics (CBE), Qatar University 5, 127–8, 131, 134, 141; curriculum 137–8, **139**; female students 135; piggybacking approach 132; relevant sustainability topics offered 137; sustainability initiative 128
corporate social responsibility (CSR) 3, 11, **12**, 14–15, 17, 52, 73, 96, 145; activities, determinants of 2; in business companies 129; certification 103; CGEM label, principles of **103**; concepts of 106, 108; departments, growth of corporations introducing 60; influential articles **18–19**; issues, urgent need to tackle 93; in Middle East higher education institutions 56; missions of business schools to incorporate 92; practices, increased need for 34; role in socio-economic development 1
corporate social responsibility (CSR) courses/program ILOs **116**; business school **109**; university degree **119**, 128
corporate sustainability: enhancing 46; global network for 58; scientific interest in integrating 129; strategic level, companies struggling to consider 129; UN Global Compact initiative 96
corruption 96, 101; CGEM principles preventing **103**; civil society actions against 103; cost of 99; in developing countries 94; efforts to eradicate 98; high levels in firms operating in Egypt 37, 39, 42; in Morocco 99, 101–2; Perception Index (CPI) **100**
corruption, fighting/working against 60, 97, 99–100, 146; importance of business ethics education in 114; Together Against strategic framework 102; *see also* anti-corruption

developing: academic achievement and skills 73; action plans 78; economies **19**, 36; ethical reflection 145; human capital 76; more ethical managers 52; multi-level frameworks 21; perception

of importance of sustainability values in daily life 146; projects 82; societies, social role of businesses in 37; teams, role of 134
developing codes: of conduct in corporations 96; of ethics for organization 61; of ethics, industry and corporate **61**
developing countries/world 3–4, 36; complexities of governance systems in 38; limited role played by governments in 37; Morocco 99; private sector as major employer 39
developing countries, CSR in 14, 17, 37; barriers 42; business ethics education in Ghana 94; conceptualization and importance of 36; contextualization of 3, 146; Egyptian case 34, 45; features differing from developed countries 56; field of study developing 4, 146; influential articles on **18–19**; in literature 38; practices 45; reflect different institutional conditions 38; religious motives 56; research domain 1; studies scarce 94; unique settings 34, 39
developing social responsibility: attitudes 77; sense of 72; among students 84
disclosure 110; corporate social and environmental 2; CSR 34, 45; corporate social responsibility **18–19**; low level of practices 42; voluntary 2
doing business: corruption as problematic factor in Morocco 99; obstacles to 39; World Bank Report 39

EBSCO host database 93
ECSRS (ethics, corporate social responsibility and sustainability) education 3–5, 8–9; business ethics 146; concepts, teaching methods preferred for understanding 138; developments in 24, 146; developments and research 10; different factors of 8; experiential learning method most preferred for teaching 141; field, overcoming micromacro divide 22; implementation on institutional level 21; increase in interest in 14; lacks in-depth insights 25; level of analysis 20; in Morocco 4; multilevel framework for analyzing 11; Nakilat integrating concepts in policies 140; new approach to learning 147; opportunities to install principles of ethical business practice in future business leaders 25; perceptions of students of proposed topic 22; prevalent concept of sustainability 22; promoting more sustainable business environments 146; research 11, 25; research in Egypt 11–12, 25, **26–8**; research output 146; shortage of pedagogical studies on 21; stand-alone course on 147; state of development of 146; state in universities 147; student attitudes and behaviors 21; studies 20; for sustainable development (ESD) highlighted 146; taking values from awareness to action 141
ECSRS (ethics, corporate social responsibility and sustainability) education integration 20; of concepts into business curriculum 138, 140; into management education, multilevel framework for 11
ECSRS (ethics, corporate social responsibility and sustainability) education in MENA region 10, 14, 22, 24; on agendas of universities 21; citations **15**; content analysis 17; ESD highlighted 146; literature in 11; sustainability as prevalent concept in 25; in universities, relevance of 147
ECSRS (ethics, corporate social responsibility and sustainability) education in Qatar 136; awareness of concepts by students 132; implementation of concepts by Nakilat 133; integration into curriculum 128, 135; at Qatar University 5; understanding how industry applies concepts *133*
ECSRS (ethics, corporate social responsibility and sustainability) literature 11, 22; articles, content analysis on 10, **12**, 13, 17, **18–19**, 24; citation-based review 145
education for sustainable development (ESD) 4; in Arab region 10; courses with emphasis on 85; in elementary school **83**; experience students in 76; GVV approach 77; higher 47, 146; human capital through 71; in MENA region 146; whole-institution approach *75*

Egypt 39; awareness of environmental, economic, cultural, and social issues 82; business environment **61**; Business Ethics course 59; cases 65; corporate elites 43; emphasis on sustainability and ethics, growth in 60; ESD 79, 84; ethical dilemmas faced by managers and employees 65; GVV approach 4, 51; higher education institutions 76, 78; importance of sustainability 46; local/multinational companies operating 41, 45; managerial attitudes towards social responsibility in *42*; SDGs 70–1; sustainable development practices 69; traffic 83; transformative agenda 70; universities participating in PRME 58; Vision 2030 71; weak regulation enforcement 45
Egypt, CSR in 44; articles on 24; barriers to implementation 42, *43*, 46; challenging environments 45; development of 5, 46; drivers in **40**; level of commitment of companies **44**; practices in 40–1; in private companies 43
Egypt, education: CSR 59, 146; ECSRS research 11–12, 25, **26–8**; gender gap 70; struggling with 70
Egyptian Agency of Partnership for Development (EAPD) 70
Egyptian business climate/environment 39–40; case 34; companies 36, 45; companies, ethical stances 42; context for CSR practices 4, 43; context, GVV cases 51, 60; market, CSR underdeveloped 45; stock exchange 40–1
El-Bassiouny, N. 3, 10, **27**, 56, 65
environmental 47, **83**; activities of corporations 34; audits/auditing 46, **136**; behavior, regulations controlling 42; biology 82; challenges, approach to **97**; considerations balanced 69; corporate performance, improving 46; damage/degradation 1, 12, 34, 73; disclosure, social and 2; education, distinct from ESD 71; fabric of the modern world, HEI role in 75; imperatives, balanced 73; improvement, barriers to 43; impact of corporate activity 45–6; knowledge 2; literacy 76; positive benefits of 71; studies, cross-disciplinary work in 77; value, sustainable **99**
environmental concerns 2, 78; efforts to raise campus awareness of 78
environmental development 59; dimensions of 70; education for sustainable **83**; policy **137**
environmental dimensions: in ESD 81, **83**; of sustainable development 70
environmentalism, value-belief-norm theory 2
environmental issues: business school has integrated concepts of 106; courses integrating main components linking 85; in Egypt, awareness of 82; integrating courses and research 58; new courses added 92; raising customer awareness on 43
environmental problems: improving 58; and potential solution 76–7
environmental responsibility 73; in Asian countries **18**; challenges in meeting **99**; initiatives to promote **97**
EQUIS accreditation 107
ethical 63; actions, students empowered to adopt 128; analysis 105–6, 131; approval of obtaining artefacts/assignments 79; challenges of today's work environment 100; choices **61**; concepts can be learned 109; conflict, resolution of 54; corporate practices 129; decisions 54, 62, 130; development of students 54; entry-level professionals 103; expectations of organizations 35; foundation for making decisions 85; frameworks 104; implementation 106, 131; knowledge of students, contributing to 56; member, action as 64; by norms and culture of the country 62; orientation by mission statements 107; principles 56; reflection 145; role play 64; scandals 51, 146; solutions 131; stances of Egyptian companies 42
ethical awareness 3–4; influenced by sustainable development 85; standards in business 112; of students 54, 85
ethical behavior 54, 73, 131; business schools mission to incorporate 92; Islamic values perceived as important guidance 57–8; provision of guidelines on 55; student 58
ethical business: decision-making, dimensions of 132; environment, Morocco's landscape in 94; situations,

ethical business: *continued*
perspective in **116**, **119**; standards in 3, 112
ethical business practices/conduct: ethics courses on, studying perceived as unimportant 55; installed by ECSRS education in future business leaders 25; in MENA countries 10; role of business schools to promulgate 8
ethical business dilemmas: choice, Muslim 57; solution to 62; students learning to deal with 131
ethical concepts; integrated into several courses 111; learned and discussed in ethics courses 109; understanding **116**
ethical conduct: business, concepts integrated into 98; courses 62, 111; misconduct 55; in Muslim business 57
ethical decision-making 52–5, 104; factors affecting 60; teaching students 103–4; techniques 4
ethical dilemmas 65; challenging/questioning values 145; confronted in the business world, correct response to 54; course focused on 52, 60; critical assessment **116**; day-to-day, decision-making in 60; GVV curriculum on 62–3, 104; implementing the right course of action when faced with 104; moral judgments in face of 113; morally reasoning out 51; Muslims faced with 56–7
ethical issues 85; analysis to understand 105; build awareness of business students 131; discussion of 56; facing Moroccan business environment 99; faculty courses covering 111; local, case studies covering 113; related to IT 115, **116–17**; student awareness of 3, 129; students understand and define 128; tools for recognizing/responding to 127
ethical managers and leaders 100; shaping 110
ethical reasoning/considerations 141; skills 110; students introduced to 131
ethical responsibilities 73, *74*, 93; of companies 41, *42*
ethical situations: analysis of **119**, 131; develop scripts for **116**; frameworks for students to use in dealing with 104
ethical values: instilled in citizens *101*; prioritizing 54
European business schools 8; lack of capacity to develop CSR education 9
European Foundation for Management Development (EFMD) 93, 107

France 94, 96
FSJES (Faculté des Sciences Juridiques, Economiques et Sociales) 106

General Confederation of Moroccan Enterprises (*La Confédération Générale des Entreprises du Maroc*, CGEM) **103**
Gentile, M.C. 1, 22, 51, 63–4, 65n1, **104–5**, 131, *133*; innovative GVV curriculum 61–2, 77–8, 84, 92, 105, 113, 128, 130, 140–1
German University in Cairo (GUC) 52, 59, 65; Business Ethics course 52, 59–60, 65
Ghana, business ethics education 94
Giving Voice to Values (GVV) 1, 4–5, 51, 59, 62, 77, 103–4, 113, 131; assumptions of **105**; case studies 61; concept 22; Egyptian cases 51; elective course 141; fosters values-driven leadership 145; founder 113; framework **116**; implemented 78; introduction to **117**; learning 131; methodology 4–5, 9, 52, 60, **61**, 62, 65; pedagogy 4–5, 65, 65n1, 141; piloted 105; premise that people already have values 77, 104, 145; program 59; reframing choices and developing action plans 78; right/ethical decision already known 62; at SBA 113–14; seven pillars of **104**; stages of 77, 82; university program 59
Giving Voice to Values (GVV) approach 1, 51, 62, 77, 94, 103–4, 113, 131, *133*, 145; action-oriented approach 132; action phase of project 139; to business ethics education 114; conceptual framework 140; encourages action to reflect own values within organization 128; flexible 131; focus on pragmatists 64; group projects as opportunity for implementing 84; implementation in a business school in the MENA region 146; importance of ethical analysis in 105–6; pedagogical, main goal of 130; positive impact of 131; skill-based 128; use in teaching 141; to values-driven leadership development 77
Giving Voice to Values (GVV) cases/case studies 63–4, **117**; challenging, critical and conceptual analysis of **61**;

developing local 113; effective 131; Egyptian suite of 51; Egyptian, use by instructors in Egypt/globally 51; focus of 63–4; local 113; move into action discussion 63; questions audience 63–4; students discuss how to get the right thing done 131; teaching through 51–2
Giving Voice to Values (GVV) curriculum 63, 92, 104–5; piloted 113; variety of business cases from different domains 62
Giving Voice to Values (GVV) methodology 52, 62, 64; for crafting scripts 106; enhances benefits of course to students 65; implementation in a business ethics course in Egypt 4; people should not be forced into acting against their values 63; strengthening human will to stand up against unethical pressure 63

higher education: accrediting bodies in 85; admission eligibility/tests 106; Association for the Advancement of Sustainability in (ASSHE) 58–9; AUI accreditation by New England Commission of (NECHE) 110; commitment from leaders 114; Moroccan 4, 94, 106; policymakers/ministries for 147; powerful tool to help create more sustainable future 140; programs, reorienting towards SD 76; role of Moroccan Ministry of 114; students 55; for sustainable development 47, 146; system, business ethics education embedded in 57
higher education institutions (HEIs) 85; business ethics courses offered within 106; business, mission statement for 107; commit to embedding principles of PRME framework 98; contribution of education for sustainable development to 4; direct impact on society 56; in Egypt, not-for-profit, private 78; experiences of students in 76; impact in 76; infusing sustainability in 75–6; integrating ECSRS education in 25; in MENA region 147; Middle East 56; mission to advance knowledge to new generations 56; in Morocco 94, 110; part of any society 74; perceived to have major role 75; teaching business ethics in 58
individual social responsibility (ISR) 73, 77, 85
institutional environment: in Egypt, hinders development of CSR 46; in many developing countries, complex 37–8; vitality in the Middle East 1
integrated/integrating: cases, delivery of 85; curricular unit 82; education for sustainable development in the Arab region 10; fourth level 21; GVV cases, into business ethics course 51; Islamic values and ethics 57; levels provide mutual reinforcement 20; marketing communications (IMC) campaign 127; perspectives of students 77; PRME principles in SBA curriculum 113; social responsibility and ethics into CBE programs 132; voices across levels of analysis 25
integrating business ethics: in coursework 56; in business curriculum 106; in programs 109
integrating CSR: in business curricula **139**; practices in Egyptian context 43
integrating ECSRS concepts: in business curriculum 132, 135–6, 138, **139**, 140; in education 11, 25
integrating ethics/ethical concepts: courses 63, 111; conversations about values and 77; and CSR concepts, student perception of 138–9; officers 96; into strategic and operational plans 21
integrating sustainability concepts: into business education curricula 129–30; into business schools' curricula 131; in curricula, student experience after 134; and environmental issues 58; in management and business education 130
integration: calls for greater 9; of ethics into organizational culture 22; ethics in universities 9; level reached 132; level of sustainability and ethics issues in business curriculum 132; proper, challenges of 129
integration of ECSRS: into business curriculum 128, 135; in management education 11
integration of sustainability: within existing structure 130, 132; level in business curriculum 132; in teaching and research 58

Intended Nationally Determined Contributions (INDCs) 70
International Association for Business and Society (IABS) 59–60
Islam 57; business ethics in **109**
Islamic: Arts and Culture 112; banks/banking sector **18**, 57; ethics **18**, 57–8; law (shar'iah) 56–7; marketing 56; values 57
Islamicization 57

Jamali, D. 1, 4, 14, 16–17, **18–19**, 24–5, 37–8, 56
Jordan 11–12, 24–5, 57, 76
Journal of Business Ethics (JBE) 14–15, 17, 25

leadership 69, **82**, 131; bottom-up approach to 112; building, students aspiring for 113; business ethics related to 51; component in PRME indicators 76; developed through extracurricular activities 109; ethical 52; GVV methodology guidance 62; models of 112; at Nakilat 137; responsibilities assigned 21; patterns 134; skills, responsible 98, **99**; university 127
leadership, values-driven 108; development 77; fostered through creative case writing 145; growing demand for 114
Leadership Development Institute (LDI) 112–13
learning environment: appropriate to support implementation of GVV concept 128; multi-pronged approach to improve 9, 20; open, for students to voice values 82; SBA 111
Lebanon: academic articles 12, 24; business schools' course curricula 59; courses being reoriented 76; CSR practice 17; ECSRS education 11, 25; education-related studies **26–8**; firms operating in 15
literature review 128; CSR 10; citation-based 4, 10, 145; systematic 10, 13, 24

mainstream/mainstreaming: approach to ECSRS concept integration in business curricula 140; approach to sustainability 130; business ethics education 94, 109; CSR education 9; CSR literature 38; CSR research 34
management 1; Academy of, SIM division 58; of change 134; commitment, top *43*, 44, 46; of company impact on society 129; Development (EFMD), European Foundation for 93; disciplines, GVV rooted in 77; ethics and **109**; ethics instruction in 3; function, criticism of 60; human resource 130; information systems 132, **136**; of information technology, ethical issues 115; *Journal of* 10; literature, macro-micro level gap 25; National Schools of Commerce and (ENCG) 106; objectives 133; perception of environmental problem 2; perceptions, study of **18**; philosophies, new 145; principles of **120**; process, student solutions on how to improve 137; professors 95; programs offered in Moroccan business schools 94; questions on 134; reference online system 12; scholarship, modern 59; stakeholder 45, 74; student ideas on 134; students interact with 140; studies, micro and macro levels 11; sustainability 73; sustainability tools 46; systems, environmental 47; systems, university 76; team capabilities **61**; Technology, GUC Faculty of 52; waste 78; Yale School of 61
management education 85; accreditation bodies advocate quality in 93; approaches to 131; Arab region, sustainable development in 10; call for change in 98; courses, introductory 108; courses and modules 132; drive to improve 98; institutions for 98; integration of ECSRS, framework for 11; Information Systems **120**, 132, **136**; integrating sustainability in 130; masters program in social solidarity economy 108; offered at schools of FSJES and business schools 106; Principles for Responsible (PRME) 8, 58, 76, 92, 98, 112, 146; promotion curriculum/course 133–4; Qatar curriculum 127, 132; research on ethics in 9; students 127, 135, **136**; sustainability topics perceived as valid by students 137
marketing 57; department, Qatar University 132; ideas proposed by students 134; integrated communications (IMC) campaign 127;

Islamic 56; objectives 133; QGBC curriculum 127; social **60**; sustainability topics covering 137
MENA (Middle East and North Africa) region 1–2, 5; countries 10–11, **12**; literature on CSR education in 1, 4
Millennium Development Goals (MDGs) 58, 69–71
Matten, D. 4, 8–9, 21–2, 34–6
Moroccan 5; citizens 101; Economy **120**; Enterprises, General Confederation of (CGEM) 103; government reforms to fight corruption 100; public university AUI 110, 112
Moroccan business environment: corruption in 99; firms, business strategies of **116**
Moroccan business schools 94; business ethics education in 100, 106–7, **109**, 114; Moroccan School of Business Administration (SBA) 110–11
Moroccan higher education: business ethics in 4; institutions 94, 98
Morocco: academic articles 11–12, 24, 24n1; anti-corruption strategy *101*; business environment ethical landscape 94; business ethics education 94, 103, 106, 114; Community Involvement Program (CIP) 111; corruption as problematic factor for doing business 99, **100**; developing country 99; ESCRS education 4, 11, 25, 146; higher education landscape 106; promotion of business integrity 101; SBA 110, 114; SBA graduates 115; unethical business behavior, attempts to eradicate 114
multinationals (MNCs) 2, 41; approaches utilized by 1; assume active roles of governance 38; managers surveyed 43; operating in Egypt 41, 45; subsidiaries, CSR of **18**
Muslims 56–7, 62–3

Nakilat Company 127; case study *133*, 134; employees 127–8, 134, 137; implementing ECSRS concepts 133, 136, 140; project 128; sustainable corporate culture 140
National Schools of Commerce and Management (*les Ecoles Nationales de Commerce et de Gestion*, ENCG) 106
New England Commission of Higher Education (NECHE) 110
North Africa *see* Middle East and North Africa (MENA)
North African countries 99

Palestine 12, **26**
piggybacking 130, 132
Principles for Responsible Management Education *see* United Nations Principles for Responsible Management Education

Qatar 4; academic articles 12; case study 140; College of Business and Economics (CBE) 127; education-related studies **26–7**; Green Building Council (QGBC) 127–8; implementation of ECSRS 5; role of ESCRS education in future business environment 146; Sustainability Week 127–8, 134; University 5, 128, 131, 141; University CBE curriculum 137
Qatar Green Building Council (QGBC) 127–8

Saudi Arabia academic articles 12, **18–19**, 24; ECSRS education research 11, 25; education-related studies 26–8
School of Business Administration (SBA), Ifrane 94, 110–15
School of Humanities and Social Sciences (SHSS) 110
School of Science and Engineering (SSE) 110
Seoudi, I. 3, 56
Siemens Integrity Initiative 113
social responsibility 35; across the curricula approach 76–7; attitudes and skills 77; AUI 111; beyond making a profit *42*; CBE focus on integrating 132; conceptual frameworks *78*; and empowerment as change agents 76; enhances cohesion and unity of society 72; towards ESD, reflection of students' 80; essential to its long-term profitability 41; exhibited by students 84; global 99; individual (ISR) 73, 77, 85; integration into university and schools mission statements 85; involvement of learners in activities related to 72; major concepts or types 73; managerial attitudes toward 41, *42*, 47n1; methods required to teach 130; practices 40, 85; practices, university commitment to 85; PRME

encouraging adoption of principles of 92; profitability and long-term goals in context of 129; programs working on developing sense of 72; proposed matrix to integrate in business education 130; reflects outcome of corporate actions 129; strategies 77; strategy behind business/managerial response to 44; of suppliers and subcontractors, promotion by corporations **103**; United Nations University involved in promoting 74; of universities 76, 80
social responsibility of businesses 35; debates over 36; knowledge of 131
social responsibility, students and: equipped with 98; perceptions of integrating 138; understanding developed 84
Social Responsibility Journal (SRJ) 15
stakeholders 36; corporate 35, 45; external communication with 46; impact of decision-making on 53; lack of effective communication channels with 45; minimum behavioral standard in relationship to 35; perceptions of ECSRS 25; PRME platform benefits for 98, **99**; PRME requires regular reporting and sharing of information among 58; profit maximization promoted over interests of 92, 97; promoting USR impacts 85–6; relevant, participation of all 70; role of employees as 43; TBL approach addresses expectations of 73; in Universities 74
stakeholders and CSR 1; affected positively by corporate behaviors 35; dimensions 3; interaction with 129; presenting plan to 45; as responsibility of company towards 40; views under-represented 146
stakeholders, involved: course provides closer look into 60; moral intensity for 53; in SBA mission 110
stand-alone courses 21, 53, 115, 130; on business ethics 65, 108–9, 114; on ECSRS 147
students: awareness of ethical issues 3–4, 129; beliefs and attitudes 9–11, 25
students' perceptions 25, 146; of ethics **19**; regarding ECSRS integration in the curriculum 128, 135, 138; of relevance of sustainability topics **136**; of SD 80
Sultanate of Oman 4; academic articles 12; education-related studies **27**; Islamicization 57; Sultan Qaboos University 4
sustainability 73, *133*; accounting tools oriented to 46; added to individual courses 130; beliefs and attitude towards 134; Bruntdland Report framework for 71; business knowledge of 131; in CBE curriculum **139**; challenges 58, **83**; co-creating **19**; culture of 46–7; Development, Center for 141; dialogue on critical issues **99**; education 52; education-related studies **26–8**; goals balanced against profit maximization 98; impact on the economy through industry 12; inclusion in MBA programs 85; incorporation in new courses 139; infusing in higher education institutions 75; initiatives implemented on campuses 141; inspired in workplace 134; of medicines 12; novel approaches used in 147; objectives 58, 132; Office of 78; policies 47; preferred learning methods for **138**; principles 92, 96; Project 59; Qatar Week 127–8, 134; routine activities related to 127; relevance of, student perception of **136**, 137; self-reporting survey (STARS) 59; systematic literature review, citation-based 145; tools, operationalization of 46–7; understanding of 136; whole-of-university approach to 76; *see also* corporate sustainability
sustainability awareness: of approaches 46; level of students in Qatar 127; raising 134
sustainability in business schools: adopting 109; new courses added 92; respond to requirements 129
sustainability concepts: added to current course structure 140; advances in knowledge and 34; background of 128–9; in CBE curriculum, student perceptions of 138; development 137; embedding in university curricula 47; included in business ethics 106; integrated in CBE at Qatar University 128; integrated in curriculum, student experience of 134; in MENA region 25; major focus of world 58; needs to be related to specific business field 130;

superseded CSR as prevalent 22; understanding **116**, **119**
sustainability, emphasis of 36; growing in Egypt 60; increased/increasing 93, 128
sustainability, global: business practice 129; develop knowledge of 60; network to promote 76
sustainability initiatives 141; by CBE 128; corporate (UN Global Compact) 96; objectives to improve students' sustainable actions 58
sustainability integrated: into business school curricula 129–30, 135; into college program 132; into common core requirements 130
sustainability issues 132; AACSB supports adoption of 52; ability of business schools to tackle 145; case study method for teaching 140; communication with stakeholders on 46; focus on 70; higher-level 37; integrating in business 130; integrating with environmental issues in courses/research 58; more established in business education 129; student awareness of 128; students exposed to 60; use of existing learning methods for teaching 131
sustainability management 73; in emerging markets **109**; tools 46
sustainability in MENA region/Middle East 25, 56, 59, 147; higher education institutions 56; mindset in 145; values incorporated in daily life 146
sustainability practice: best-known 58; development and implementation of 85; global business 129
Sustainability Tracking, Assessment & Rating System (STARS) 59
sustainable development (SD) 1, 74, 80–1: ASSHE aims to make norm among universities 58; business school courses in **109**; culture, cultivation of 46; Decade for (DESD) 59; in Egypt, gap related to 84; Egyptian HEI that caters for 78; Egypt's strategy for 70; Goals (SDGs) 70; grading system needed 80; important issue for discussion 34; infusing 76; investment to achieve 70; learning for/as *72*; link with education 71; performance reporting on 73; practices in the Arab region 69; reformed learning as form of 72; reorienting higher education programs and courses towards 76; role of academia in promoting tool to promote 46; student perceptions of 80; student perception/understanding of the concept 80, **136**, 146; three dimensions of 70; UAE University research centers focusing on 59; universities driven by three realms of 74; underpinning principles of 8; way to improve environmental problems 58; *see also* education for sustainable development
Sustainable Development Goals (SDGs) 70–1, 76, 84
Syria 11–12, 24–5

Tormo-Carbó, G. 3, 85
Trans-European Mobility Scheme for University Studies (TEMPUS)-funded project 76
Transparency International (TI) 99–100; global strategic framework Together Against Corruption *102*
Transparency Maroc 101–2
triple-bottom-line (TBL) 73

UK 76, 94
UNESCO (United Nations Educational, Scientific and Cultural Organization (UNESCO) *75*; Decade for Sustainable Development) 69–71
unethical behavior 104; anticipate rationalizations for **104**; daily 55; education and training in fight against 101
unethical business conduct 62; dealing with *101*; hindering development 100; business conduct in Morocco 99, 114; in the workplace 55
unethical business practices: employee pressured into 62; international cases of 60; unprecedented levels of 8
unethical: opposition 64; pressure, training to stand up against 63; use of data alleged **93**
United Arab Emirates American University of Sharjah, UAE 58
United Arab Emirates (UAE) 12; American University of Sharjah 58; CSR and sustainability education 59; ECSRS education research 11, 25, **26–7**; education research study 57
United Nation Principles for Responsible Management Education

(UN-PRME) 8, 58, 76, 92, 98; approach to business ethics education 114–15; indicators/areas 76; framework principles **99**; launched by UN 97–8; principles 58, 98, 113; role of 94; signatories 8, 59, 98, 112–13
United Nations Convention against Corruption 100
United Nations Framework Convention on Climate Change (UNFCCC) 70
United Nations Global Compact 92, 96, **99**; Leaders' Summit 98; ten principles **97**
United Nations Principles of Responsible Management Education (UN-PRME) 8, 58–9, 76, 92, 94, 98, 112, 114, 146; principles 58, 98, **99**, 113; signatories 8, 98, 112–13; website 98, 113
United States (US/USA) **26**; Sarbanes-Oxley Act 96; scandals 94, 96
United States (US) business schools 8, 95; courses teaching business ethics in business curricula 97; Darden School of Business 59
university social responsibility (USR) 73, 75, 77, 84–5

Vision 2030 strategy for sustainable development, Egypt 70–1, 84

WE garden **82**
Windsor, D. 73, 145
World Bank 39–40; Doing Business 2016 Report 39
World Commission on Environment and Development (WCED) 71

Printed in the United States
by Baker & Taylor Publisher Services